# In Ohio's Backyard: Periodical Cicadas

by
Gene Kritsky, Ph.D.

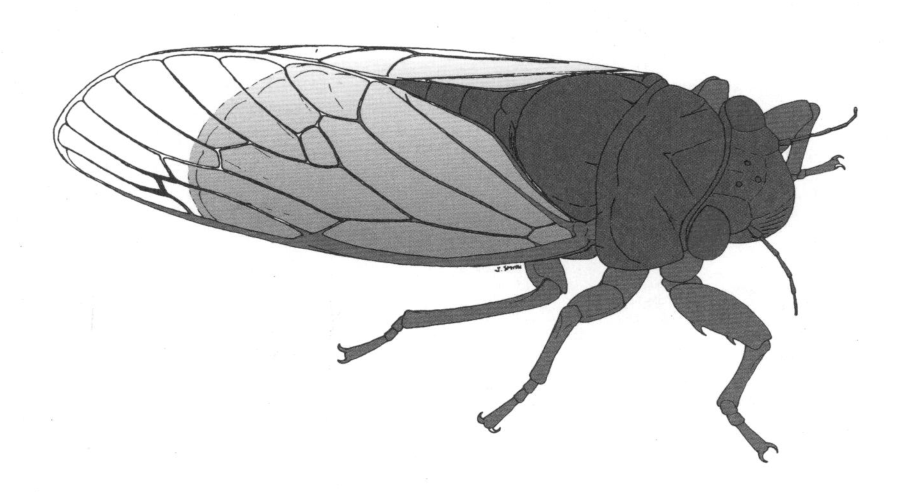

**Ohio Biological Survey**
**Backyard Series No. 2**



**1999**

Drawings by Ann E. Geise

# Ohio Biological Survey

Backyard Series: ISSN: 1095-8800
Backyard Number 2 ISBN: 0-86727-132-9
Library of Congress Number: 99-070680

Brian J. Armitage, Director
Veda M. Cafazzo, Editor

**Cover Photograph**



**Distributor**

Ohio Biological Survey, 1315 Kinnear Road, Columbus, Ohio 43212-1192 U.S.A.

**Copyright**



**Literature Citation**

**Kritsky, Gene. 1999.** In Ohio's Backyard: Periodical Cicadas. Ohio Biological Survey Backyard Series No. 2. Columbus, Ohio. vi + 83 pp.

**Credits**

Layout and Design: Brian J. Armitage, Ohio Biological Survey
Illustrations: Ann E. Geise, Cincinnati, Ohio
Printing: Ohio State Univ. Printing Services, Columbus, Ohio

Ohio Biological Survey
College of Biological Sciences
The Ohio State University
Columbus, Ohio 43212-1192
http://www-obs.biosci.ohio-state.edu

**5-99—5M**

# Preface

No insect attracts public attention like the periodical cicadas. They were the first insects to be noticed in the New World—and today, when they emerge, they receive national press coverage. This is because most insects live generally secret lives, but not the periodical cicadas. When they appear they do so as an invasion. After their arrival they sing so loudly that you cannot ignore them and then they are gone for 17 years. But in reality they are not gone, for millions are living under trees all over Ohio and much of the eastern United States. In 1999 they will appear in Cleveland and most of eastern Ohio. In 2004 they will inundate Cincinnati and much of southwestern Ohio. Indeed, within the next decade, most parts of Ohio will be visited by this wonder of nature and history. To truly understand periodical cicadas you must know their history. Because it is their history that explains the myths that surround these insects. After you know their history, then you can better appreciate their biology.

The goal of this book is to be the starting point on your discovery of these insects. The first task is to introduce the history of our curiosity about these insects and how scientists used numbering systems to deal with the confusing life cycle. Then we will examine Ohio's periodical cicadas by detailing each of our broods. You will also learn about the periodical cicada's biology, morphology, and behavior. Throughout the book you will be presented with projects that you can do to appreciate these insects better. Following a glossary to handle any technical terms, you will find a series of appendices. These include tables for Ohio's historical cicada emergence records organized by county, emergence tables for past and future cicada broods, and distribution maps for all U.S. broods. Armed with this information readers can determine if periodical cicadas have appeared in their areas and when they are expected again. Finally, a bibliography covering over 350 years of interest in and research on periodical cicadas is included as the final appendix.

People complain about the multitude of periodical cicadas when they appear, but they are as unique to Ohio and the eastern United States as giraffes are to Africa. Come and explore an insect that may very likely be living under your backyard!

Acknowledgements

Many people have helped me with my study of the periodical cicadas and this work is, in part, a result of the many conversations I have had with them. I would like to thank Monte Lloyd, Chris Simon, and Thomas E. Moore for their help in this regard. Three periodical cicada specialists who greatly influenced my interest in cicadas are no longer with us to enjoy these insects. Frank N. Young, Jr. first sparked my interest in periodical cicadas when I was his undergraduate student and I am proud to have co-authored several papers with him. Henry S. Dybas and Lewis J. Stannard, Jr. were also pivotal to my early work.

I also thank Susan Reidel who read over this manuscript, produced the national maps, provided photographs, and endured hours of travel in search of periodical cicadas. Thank yous also go to Jessee Smith who verified the historical records and provided the line-art drawing of the periodical cicada, Sue Simon who helped to compile the historical records and reported the unexpected emergence in 1995, and to Matthew Hohmann who provided striking photographs taken during the 1995 emergence in Athens. Unattributed photographs in this book are by the author.

I also want to thank my colleagues at the College of Mount St. Joseph, Annette Muckerheide, Elizabeth A. Murray, Kathleen S. Prezbindowski, Mary Schilling, Richard A. Davis, and Marlene Pohlmann, for their support and help especially when I received literally thousands of calls regarding the periodical cicada emergences.

Several grants have provided support for my periodical cicada work and I thank the College of Mount St. Joseph for research grants and a sabbatical leave. I am indebted to the Cincinnati Museum of Natural History and Science—Collections and Research for support of my work as an Adjunct Curator. I also thank the Ohio Department of Natural Resources, Division of Wildlife for funds to study Ohio's Brood XIV and funding of the "In Ohio's Backyard" series. I thank Veda M. Cafazzo for her care in editing this work. I am grateful to Susan Adkinson, Carolyn Caldwell, Jan Hall, Valerie Jacobs, and Dr. Charles A. Triplehorn for their thoughtful and detailed review of the manuscript. Finally, I am especially indebted to Brian J. Armitage and the Ohio Biological Survey for support and encouragement of this and my other projects.

This book is dedicated to Dr. Frank N. Young, Jr. for his teaching, enthusiasm, and love of periodical cicadas.

Gene Kritsky
Cincinnati, Ohio

# Table of Contents

# List of Maps

# CHAPTER 1. A Bug of History

It occurred in 1633, just twelve years after the first Thanksgiving at Plymouth. Some kind of insect appeared in incredible numbers. William Bradford, the Governor of Plymouth Colony wrote:

> And the spring before, especially all the month of May, there was such a quantity of a great sort of flies like for bigness to wasps or bumblebees, which came out of holes in the ground and replenished all the woods, and ate the green things, and made such a constant yelling noise as made all the woods ring of them, and ready to deaf the hearers. They have not by the English been heard or seen before, or since. But the Indians told them that sickness would follow, and so it did in June, July, August and the chief heat of summer.

What the pilgrims witnessed was an emergence of the periodical cicada. What made the event noteworthy was it did not happen again for years. Indeed, it was well over a decade before anything like it was seen again. The account of this insect plague was repeated in an English publication 33 years later in 1666 by Henry Oldenberg who published a paper titled: "Some Observations of Swarms of Strange Insects, and the Mischiefs done by them."

Oldenberg wrote:

> A great Observer, who hath lived long in New England, did upon occasion, relate to a Friend of his in London, were he lately was, That some few Years since there was such a swarm of a certain sort of Insects in that English Colony, that for the space of 200 Miles they poysion'd and destroyed all the Trees of that Country; there being round innumerable little holes in the ground, out of which those Insects broke forth in the form of Maggots, which turned into Flyes that had a kind of taile or sting, which they struck into the Tree, and thereby envenomed and killed it.
>
> The like Plague is said to happen frequently in the Country of the Cosacks or Ukrani, where in dry Sommers they are infested with such swarms of Locusts, driven whither by an Haft, or South-East Wind, that they darken the Air in the fairest weather, and devour all the Corn in that Country; laying their Eggs in Autumn, and then dying; but the Eggs, of which every one layeth two or three hundred, hatching the next Spring, produces again such a number of Locusts, that they do far more mischief that afore, unless Rains do fall, which kill both Eggs and Insects themselves, or unless a strong North or North-West Wind arise, which drives them into the Euxin.Sea:[sic] The Hogs of that Country loving these Eggs, devour also great quantities of them, and thereby help to porge the Land of them; which is often so molested by this Vermine, that they enter; into their Houses and Beds, fall upon their Tables and into their Meat, insomuch that they can hardly eat without taking down some of them, in the Night when they repose

> themselves upon the ground, they cover it three or four Inches thick, and if a Wheel pass over them, they omit a stench hardly to be endured.

The story linked the strange insects from Plymouth with a locust plague and so began an error that lives on today. Are these strange insects locusts or cicadas? The English colonists thought the insects were a plague like the locusts of the Bible. Moreover, the Bible recorded that John the Baptist ate locusts and the Native Americans ate these insects also. Therefore, they must be locusts. But the locusts of the Bible were grasshoppers and the insects that were seen in Plymouth Colony did not look anything like grasshoppers. They were named *Cicada* by Carolus Linnaeus in 1759, but this scientific name given to these insects by the Swedish scientist had little impact on the English and millions still call them locusts to this day.

Other legends started to surround these insects. By the middle 1700s people in North Carolina started claiming that the cicada could predict war. In 1764 in Dobbs Parish it was recorded about the cicadas:

> The country people are superstitious about them, and when the Locusts come they look on their wings to see whether war or a good time is predicted. There are usually black lines on their thin wings, and the people read them as letters, and say that W means war, P means Peace, and so on.

So common was war during this period before the American Revolution that the cicada always predicted war. In fact, that is all they can predict because the W is formed from wing veins and they are always in the form of a W.

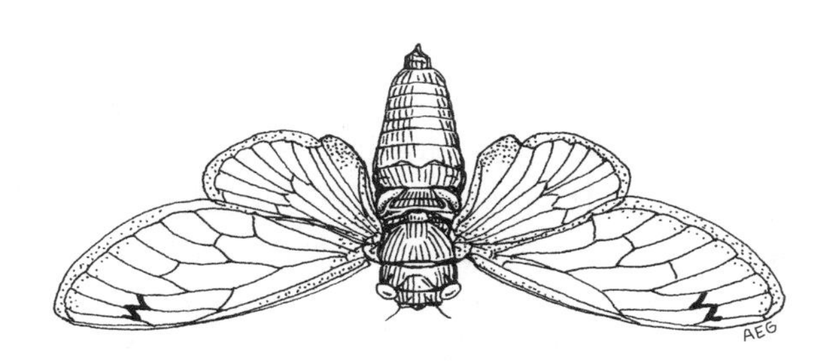

One hundred and fifty years after the cicadas were first observed the periodical nature of their life cycle was suspected. But the duration of the life cycle was not at all clear. Some regions had more than one population of cicadas each appearing every 17 years, but the overlapping populations created confusion about the length of the life cycle. Many parts of the east had two populations that appeared four years apart. However, to the observer it appeared that they were the same population that occurred at irregular periods. For example, in 1715 the cicadas appeared in Philadelphia, and they did so again four years later. This was followed by another emergence 13 years later still, which was followed by cicadas in another four years. Other parts of the east had cicadas appearing every 17 years. It was after decades of observations that it was determined that the cicadas had a 17-year life cycle and those areas with cicadas appearing more often than 17 years had more than one population of cicadas!

It was Dr. S. P. Hildreth who lived in Marietta, Ohio who helped confirm that the cicadas indeed had a 17-year life cycle. He wrote a detailed account of the 1812 emergence of the cicada and verified that they previously appeared in that part of Ohio in 1795. He followed up this paper with a second paper describing the emergence of 1829. His observations provided documentation of three consecutive 17-year emergences of a single population.

Adding to the confusion concerning their life-cycle was the discovery that some of the cicadas in southern states had a 13-year life cycle. This was first proposed in 1845 when Dr. D. L. Phares wrote in the Woodland, Mississippi *Republican* that the cicadas had emerged there after 13 years. He reconfirmed his observations in 1858 when they emerged again.

This confusing array of periodical cicadas needed order. Several systems were proposed but the one that is in use today was proposed in 1898 by a federal bureaucrat, Charles L. Marlatt. His system was a simple numbering of the years cicadas emerged. He designated those cicadas that emerged in 1893 as Brood I cicadas. The cicadas that emerged in 1894 were called Brood II. The cicadas emerging in 1895 were to be called Brood III and so on. Because there were 17 possible emergence years, numbers 1 through 17 were reserved for the 17-year periodical cicadas. Numbers 18 through 30 were to be used for the possible 13-year cicadas. Although there are 30 possible broods, there are in fact only 12 established 17-year broods and only three established 13-year cicada broods. Ohio has four established broods. Using Marlatt's numbering system Ohio's broods are designated V, VIII, X, and XIV.

Marlatt's system greatly reduced the confusion surrounding the study of periodical cicadas and it is still used today over a century after it was proposed. It helped in areas where there were overlapping broods to determine when cicadas would again emerge.

Until 1940, the United States Department of Agriculture continued to take care to map out the broods of periodical cicadas. Their efforts were recorded in stenciled and stapled reports that included all the historical data as well as where cicadas had emerged in the current year. With the outbreak of World War II this organized mapping effort stopped and never started up again.

The national mapping effort provided good distributional maps of the periodical cicada broods. The map below indicates the distribution of all of the 17-year (blue dots) and 13-year (red dots) broods. This map shows that the 17-year broods are distributed north and west of the 13-year broods. Maps found in Appendix C show the national distribution of the individual broods.

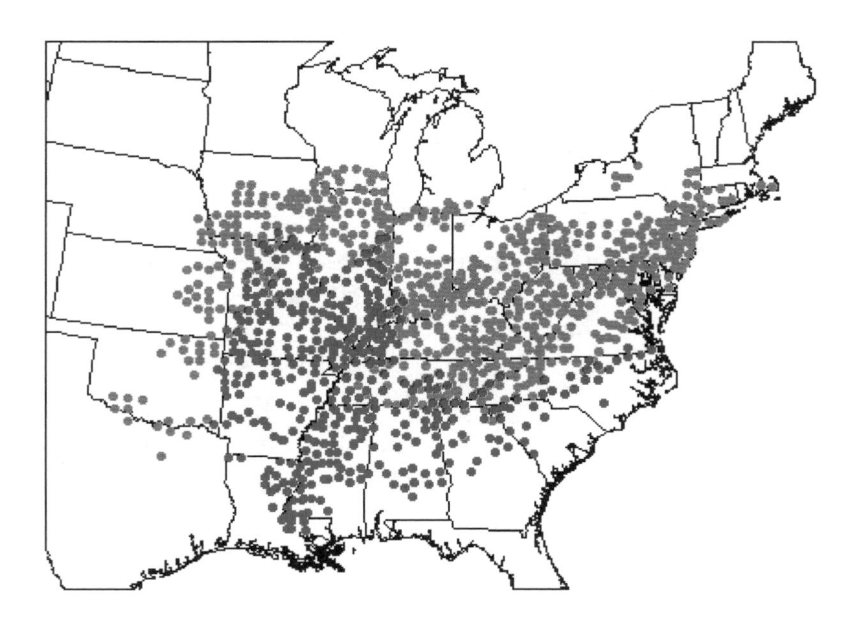

Distribution of 13-year (red dots) and 17-year (blue dots) cicada broods in the U.S.

The USDA records for Ohio and other published historical accounts have been compiled into an appendix included in this book. To determine if your area is home to a periodical cicada brood, simply look up your county and town. If cicadas have emerged there, it is listed along with the date of the emergence. The brood number is given in Roman numerals following the system proposed by Marlatt. To find out when the brood is next expected, look up the brood number on the Emergence Table and scan across to a future year.

Given the decline of the periodical cicada across its range, many of the historical dates may represent emergences that no longer occur. After you have determined that your area has had emergence, compare the information with the current distribution maps to see if they are still expected. If the cicadas are no longer expected, then you can help with the mapping efforts. Every emergence that has been recently studied has produced many new emergence records. If you find periodical cicadas in a part of the state that is not included in the appendix of historical records, then please send that information to the author for inclusion in new maps.

## Suggested Readings

For more information on the early history of periodical cicadas see Marlatt (1907).

### Projects

Look up your area in the appendix to determine if there has been an emergence in your town. If there has been an emergence, then go to your local library or historical society and search the May and June issues of the local newspaper for the emergence year. Did the paper record the appearance of the periodical cicadas? Copy and assemble a poster of earlier accounts of local emergences.

If your area did not have an emergence, then look up the emergence years for New York or a large city that had a newspaper that your library has on microfilm. Check the May and June issues for the emergence years. How far back can you find records of cicada emergences?

# Chapter 2. Ohio's Broods

## Brood V

The Buckeye State is home to four established broods of 17-year cicadas. These broods cover every part of the state although some counties have seen their periodical cicada populations disappear. The largest eastern brood is Brood V which will next occur in 1999 and every 17 years thereafter (map on next page). Brood V was first recorded in Ohio in 1795 and was described in detail in 1812 by Dr. S.P. Hildreth who lived in Marietta. His description is typical of these early observations detailing the emergence of the adults and is included at the end of this chapter.

Brood V was first mapped out by Francis Marion Webster who was working for the Ohio Experiment Station. Webster's map of the 1897 emergence was published in November of that year and showed the brood to occur in nearly every county of the eastern half of the state. Typical of scientific writing of the day, his paper on the periodical cicadas of Ohio included large sections lifted from earlier published writings. He also included transcripts of all of his correspondence with Ohioans about the cicadas.

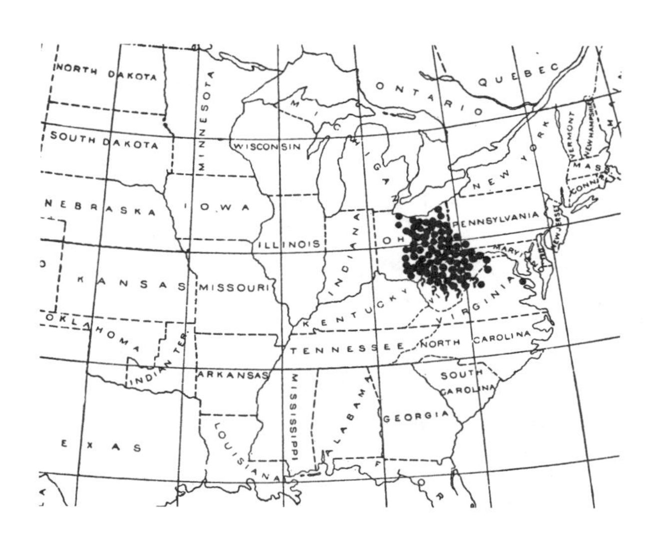


Marlatt's 1907 Brood V map.

In 1907, Charles L. Marlatt, acting Chief of the Bureau of Entomology at the United States Department of Agriculture, published a bulletin on the periodical cicada including maps of all the broods. His map for Brood V was based on Webster's map and information from letters received from state entomologists, school superintendents, and mail carriers. Letters he received from state entomologists reveal the difficulties in determining the cicadas' distributions in a time when transportation was slower and communication more difficult than it is today.

The USDA continued to collect distributional data on cicadas through 1940. T. H. Parks of The Ohio State University mapped out the 1931 emergence of Brood V and forwarded his information to the federal government. The last time Brood V was carefully mapped was in 1976 when H. Y. Forsythe, Jr. studied the brood in Ohio. His map showed only minor differences from earlier maps. Plans are already being made to make detailed maps of the 1999 emergence of Brood V in an effort to determine if the brood is showing a decline similar to the other Ohio broods.

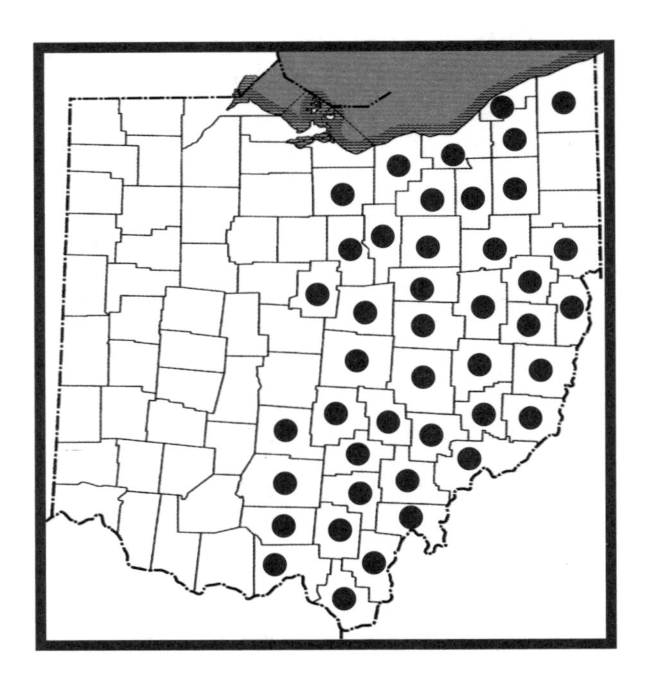

Map of the Ohio distribution of Brood V. In maps to follow, circle symbols will vary in size to indicate heavier emergences (larger diameter) or lighter emergences (smaller diameter) in cicada populations in a given area over time.

## Brood VIII

The smallest brood in Ohio is Brood VIII which is confined to just six counties in extreme eastern Ohio (map below). Dr. Gideon B. Smith in Baltimore, Maryland, in a manuscript now lost, reported that Brood VIII occurred in Columbiana County in 1815, 1832, and 1849. This observation was documentation that Brood VIII overlapped with part of Brood V which emerges three years ahead of Brood VIII.

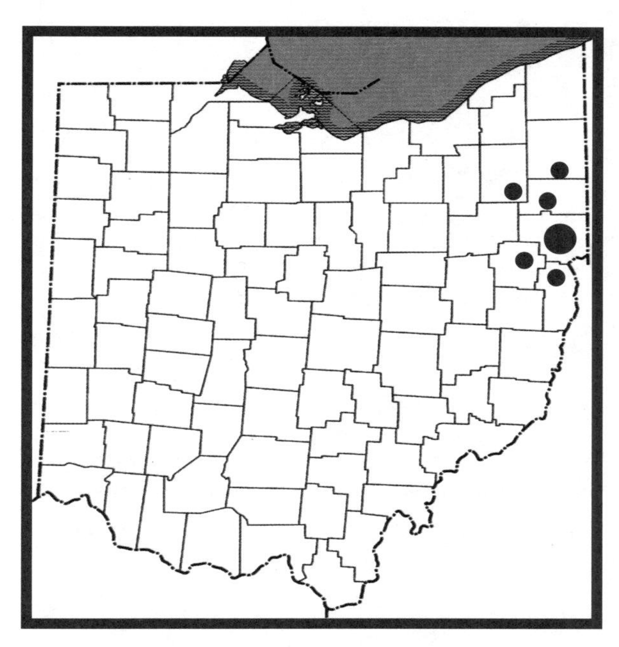

Map of the Ohio distribution of Brood VIII.

## Brood X

Brood X is the largest of the 17-year cicada broods extending across much of the eastern United States. In Ohio, Brood X is found in the western half of the state with the heaviest emergence taking place in Hamilton, Butler, Montgomery, Clark, and Logan counties (map on next page).

Because Brood X occurs over such a large part of the country it has attracted considerable attention. In 1868, a front page story in the *Cincinnati Enquirer* started with the following: "On a ride through one of the rural suburbs of our city yesterday, our ears were greeted with the singing, as it were, of myriad's of locusts; and, occasionally, so powerful was the noise that conversation was interrupted."

The oldest record of Brood X in Ohio dates from 1817 in Clermont County. Since then its appearance is documented in newspaper accounts during subsequent emergence years.

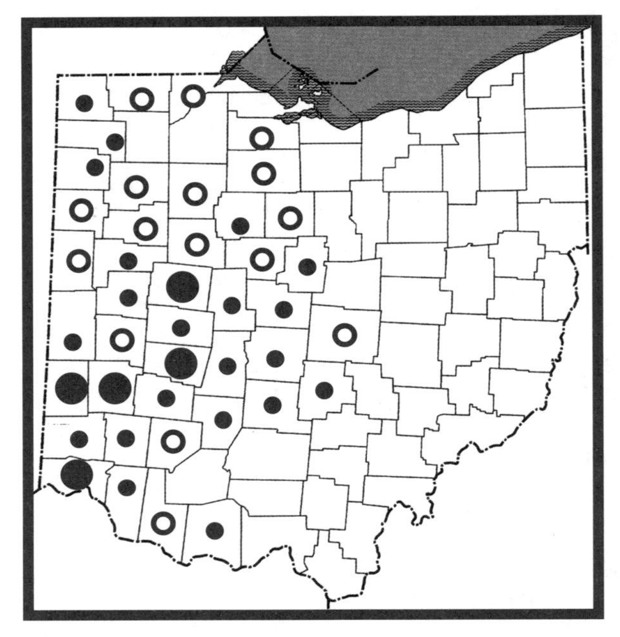

Map of the Ohio distribution of Brood X. Open circles represent counties where Brood X has been recorded, but did not emerge in 1987.

## Brood XIV

The last brood that occurs in Ohio is Brood XIV. This brood occurs in the southern portion of the state with sporadic individuals found as far east as Washington County. Like Brood X, this brood has a wide distribution over much of the eastern United States. The oldest record of this brood in Ohio is 1804 from Brown County.

Interactions between Broods X and XIV can be seen in Cincinnati. In 1906, Brood XIV emerged in western Cincinnati in Delhi and Western Hills. In 1991, only about 20 adult cicadas emerged on the West Side, whereas millions emerged there in 1987 when Brood X emerged. A careful mapping of Brood XIV shows that it has slowly been pushed eastward by Brood X. In 1991 nearly all of the Brood XIV occurred in the easternmost suburbs of Cincinnati.

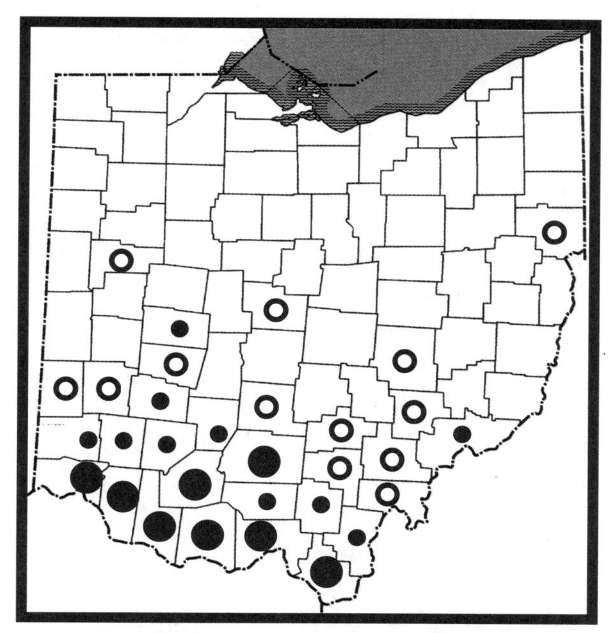

Map of the Ohio distribution of Brood XIV.

## Unexpected emergences

Periodical cicadas are not always punctual. Oftentimes they emerge four years ahead of schedule and other times thousands will emerge one year late and these unexpected emergences have occurred in Ohio. Our citizens have usually taken notice of the periodical cicadas even when they are unexpected and that notice has provided documentation of these "exceptions" and is yielding valuable clues to the evolution of the broods.

The first documented unexpected occurrence of the periodical cicada in Ohio took place at Spring Grove Cemetery in Cincinnati in 1898 four years ahead of the expected Brood X. Accelerations also took place in Cincinnati in 1932, 1966, and 1983. In 1995, a four-year acceleration of Brood V occurred in Athens, Ohio as well as smaller emergences in Hocking, Washington, and Gallia counties.

In 1988, Cincinnati witnessed a major one-year deceleration from Brood X with thousands of cicadas emerging in Woodland Mound Park. In Northern Kentucky, across from Cincinnati, the emergence was even heavier. Thousands of periodical cicadas emerged in California and Alexandria.

These unexpected emergences show how the new broods evolve. It appears that as broods produce large populations some individuals in the population will emerge four years ahead of schedule. This would result in a heavy population slowly declining as another brood that emerges four years earlier, becomes established, and builds its population. As that brood increases it too will have four-year accelerations. That is precisely what is happening in Ohio and Indiana where the periodical cicadas have been mapped for over a century. Apparently Brood XIV has been accelerating into X during the last century and X is showing accelerations this century that would be designated Brood VI if the accelerated populations become established. This explains why many neighboring broods are four years apart in their emergence years.

In eastern Ohio, the relationship of Broods V and VIII suggests a slightly different process. Brood V and VIII are three years apart and a combination of a four-year acceleration combined with a one-year deceleration would produce a three-year difference between neighboring broods.

## The decline of Ohio's broods

To a homeowner inundated with thousands of screaming cicadas, this may sound strange, but we are seeing a decline in periodical cicadas! In fact, an ecology textbook even suggested that the periodical cicada is threatened. The decline is likely due to agricultural practices and clear-cutting for development. The distribution maps for Broods X and XIV tell the tale for Ohio. The open circles are areas where the cicadas have emerged in the past but no longer occur and the smaller circles represent counties where the last emergence was light. The clear message is that periodical cicadas are not as widespread as in the past. In Indiana the decline is even more apparent with an approximate 40% reduction in Brood X's distribution.

This decline seems contradictory to people living in urban areas where the periodical cicadas are abundant. That is because periodical cicadas are becoming more common in cities and towns and less abundant in undeveloped areas. This is caused by the female cicadas' choice of trees in which to lay their eggs. To insure that the tree will survive for 17 years to feed the young, female cicadas seem to select young trees in full sunlight surrounded by low vegetation. This is a wonderful description of a front yard or a city park.

Ideal location from a cicada's point of view.

## Hildreth's account of the 1812 emergence in Ohio

Our insects are so abundant and so various, that it would take a volume to describe them alone. One of the most interesting and curious of this class is the Cicada. It nearly resembles the harvest-fly, but is smaller. They are said to appear only at stated periods, which some have fixed at seventeen, and others at fourteen years. I have one record of their appearing in this country, the 14th of May, 1812. I was then told it was seventeen years since they were last here, *viz.* in 1795. They gradually disappeared, and by the first of July were all gone. The month of May was cold and wet, and very unfavourable to the egress of the cicada from the earth. From the 24th of May, to the 3d of June, their numbers increased daily, at an astonishing rate. The cicada, or "*locust*," as he is vulgarly called, when he first rises from the earth, is about an inch and a half in length, and one third of an inch in thickness. While making his way to the surface, he has the appearance of a large worm or grub; the hole which he makes is about the same diameter with his body, perpendicular, and seems to be made with equal ease through the hardest clay or softest mould. When they first rise from the earth, which is invariably in the night, they are white and soft. They then attach themselves to some bush, tree, or post, and wait until the action of the air has dried the shell with which they are enveloped: the shell then bursts on the back for about one third of its length, and through this opening the cicada creeps, as from a prison. Their bodies are then very tender, and they can neither fly nor crawl to any considerable distance. In this state they remain until morning, their wings gradually unfolding, and as the day increases, they, by little and little, and frequent attempts, learn to fly for a few feet, so that by night they are able to fly for several rods. In their efforts to disengage themselves from their shell or envelope, I noticed that many of them lost their lives - either from a want of strength to burst away, or from the narrowness of the passage, occasioned by their coming to the surface of the ground too early, and the action of the air drying, burst their covering before their bodies were prepared for the change. In a diary which I kept at the time I find the following observations.

*May 27.* - I find the following record. "This day, and for two or three days past, the locust, or cicada is beginning to appear in vast quantities on the trees and bushes in the woods; they seem yet not to be fully grown, nor very active, but are easily caught. The hogs are very fond of them and devour all they can find, and indeed they seem to have commenced their attack upon them, by rooting, before they left the ground. It is thirteen days since they first began to break from the earth, but did not leave their holes in any great numbers, on account of the cold, till lately." The last of June, the cicadæ gradually disappeared. At this time the females were very weak and exhausted; and some which I examined, appeared to have wasted away to mere skeletons, nothing remaining but their wings and an empty shell of a body. Since that time few, or none, have appeared in this county; but I have heard of their being seen in some of the neighbouring states, I believe east of the mountains.

*June 3.* - Yesterday the cicadæ were seen making preparation to lay their eggs.

*June 4.* - The cicadæ begin to deposit their eggs in the tender branches of apple-trees: they appear to be very fond of young trees of this kind, and of the forest-trees they seem to have a decided preference for the beech, on which they collect in vast multitudes; and when any one passes near, they make a great noise, and screaming, with their air-bladders, or bagpipes. These bags are placed under, and rather behind the wings, in the axilla, something in the manner of using the bagpipes, with the bags under the arms - I could compare them to nothing else; and indeed I suspect the first inventor of the instrument borrowed his ideas from some insect of this kind. They play a variety of notes and sounds, one of which nearly imitates the scream of the tree-toad.

*June 12.* - The cicadæ still very busy depositing their eggs in the tender branches - which branches die and fall off. The male only makes the singing noise from the bladder under his wings. The female has no wind instrument, but an instrument like a drill or punch, in the centre of her abdomen with which she forms the holes to deposit her eggs - the same instrument also deposits an egg at the instant the hole is made. The punctures, or holes are about an eighth of an inch apart, and in the heart or pith of the branch on its under side. One cicada will lay an immense number; by the appearance of one I opened today each fly is furnished with at least one thousand eggs.

While the cicadæ remained with us, I could not discover that they made use of any kind of food, although I examined them repeatedly and particularly for this purpose. All the injury they did to vegetation, was in depositing their eggs; by this process they materially injured, and in some instances nearly destroyed, young orchards of apple-trees. Many of them to this day will bear ample testimony to the truth of this remark, in their mutilated limbs and knotted branches.

In addition to the foregoing observations, I have learnt to a certainty, that it is seventeen years since the cidadæ were here before. Early in the spring of 1795, a clearing was commenced eight miles above this place, on the Muskingum, and an orchard put out on the piece, perhaps half an acre, that was cut over before the cicadæ appeared & the rest of the clearing was made the same season, after they had disappeared. When they again appeared in 1812, it was observed by Mr. Wright, the occupant of the land, that not one cicada came out of the earth on that piece of ground where he had cut the trees before they appeared in 1795; but that on all the rest of the land, wherever there was a stump, or a tree had stood, the earth was full of holes made by the ascending cicadæ. These facts are in my mind a sufficient evidence that it is seventeen years between the laying of the egg, and the re-appearance of the cicada. Through how many transformations they pass, is to me unknown; but from the length of time they lie in the earth, it is probable the changes are more than one. But that they do not travel far is evident, from their coming up immediately by or under the spot, where the tree stood in which the eggs were deposited.

## Suggested readings

For detailed information on Ohio's broods see Forsythe (1976c), Kritsky (1988a), Kritsky (1992), and Kritsky and Simon (1996). For details about the evolution of the broods see Kritsky (1988b) and Williams and Simon (1995).

## Project

If you are in the midst of a periodical cicada emergence, map out the local distribution and keep a diary similar to Hildreth's 1812 diary describing the weather, when the cicadas emerged, started singing, laid eggs, and died. What trees are attractive to cicadas? Are cicadas more common in certain areas?

# Chapter 3. The life of a periodical cicada

The emergence of a periodical cicada from the ground is often the first time people notice these insects. But this is not the start of the periodical cicada's life but rather near the end of that individual's existence. The emerging nymph is actually the last immature stage of the cicada's life. That same evening the cicada has emerged from its darkened tunnel, it will climb an upright surface of a tree, fence, or wall; anchor its feet into the surface and transform into an adult. This transformation is a wonder of nature. The immature cicada will split its nymphal skin at its back. This split will grow revealing a milky white adult. As the split becomes larger, the adult will pull itself out of the old skin in a seemingly slow-motion acrobatic process. First, the head and thorax of the cicada are pulled out. This is followed by the legs and the remains of its breathing tubes which are called trachea. With only the narrow opening of the old skin holding the insect's abdomen, it will pull its body out and nearly flip upside down. The white adult with its red eyes then does a "sit-up," grabbing the old skin with its legs and enabling it to free its abdomen.

Adult doing a sit-up.

Photo: Susan Reidel

At this time the wings are thick little flaps that have to be "inflated" like a balloon. Fluid is pumped into the wings forcing them to expand. After the wings have expanded, the insect begins to darken and harden. Finally, the cicada will appear black with red eyes and orange-red wings. Now it is a fully formed adult and is ready to find a mate.

Typical emergence sequence for periodical cicadas. A. holes in soil where nymphs have emerged; B. nymphs climbing up onto a tree trunk; C-E. adult emerging from the nymphal skin (note unexpanded wings); F. fully emerged but teneral (soft exoskeleton) adult extending wings; G. teneral adult with fully extended wings; H. mature adult with hardened exoskeleton. [Photos: A and B - Susan Reidel; C-H - Matthew Hohmann]

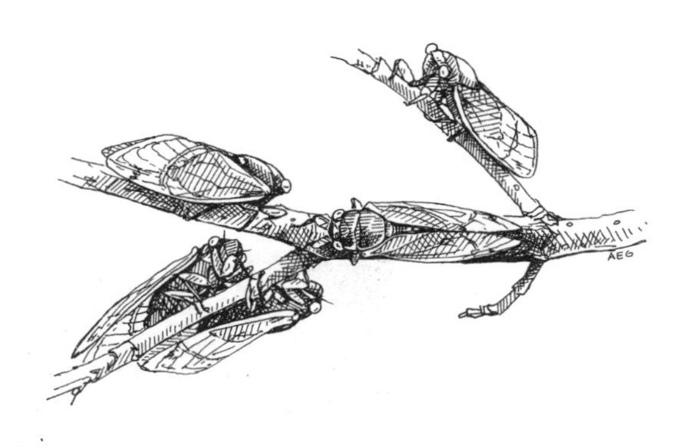

Chorusing males in tree branch.

Four to five days after their emergence, the male cicadas start to sing. Males will gather in trees where they sing together to attract mates. These collections of singing males are called chorusing centers. When a female flies into the tree she apparently selects a mate in part by size, favoring larger males over smaller ones.

After mating, the female will lay her eggs in the twigs of trees. The choice of a tree is a critical one for the female because she must find a tree that will live for at least 17 years to insure that the developing young will have tree roots on which to feed. If the tree should die or be cut down, the young cicadas could perish. For this reason, female cicadas appear to select smaller trees surrounded by low vegetation and in full sunlight. Once a tree is selected, she uses her serrated ovipositor to cut into the new growth of the tree which is the very end of the branches. This cutting of the tree creates a scar on the twig that can be seen on the tree for years. Once the ovipositor is inserted into the twig, she then deposits between 10 to 30 eggs until all of her approximately 400-600 eggs are laid.

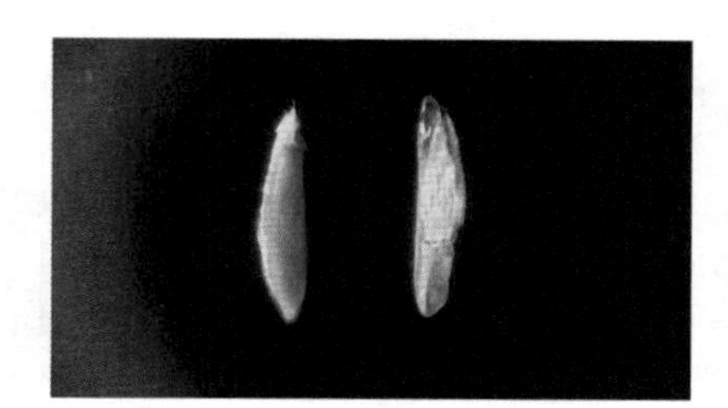

Hatched and unhatched cicada eggs.

Oviposition scar and cutaway of eggs.
[After Riley, 1869]

After about a month of singing and sex, both the male and female cicadas die. Their bodies collect at the bases of trees where they decay in the warm summer sun. The odor of millions of rotting cicadas is sometimes as overwhelming as the emergence.

Dead cicadas at the base of a tree.

After six to eight weeks the eggs start to hatch and microscopic cicada nymphs leave the twigs and fall to the ground. They apparently feed on grass roots for a time and then dig down 12 to 16 inches where they begin a dark life of tunneling around the tree roots feeding on juices from the roots. The cicadas grow slowly and at different rates, molting into larger nymphs. The cicadas are not sleeping while they are out of sight. They are busy digging tunnels and feeding.

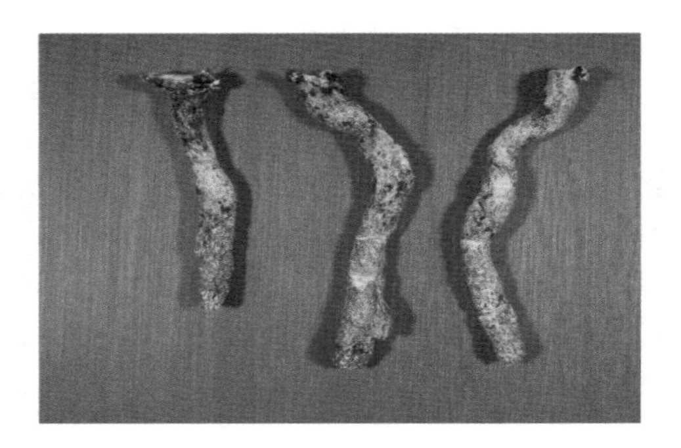
Plaster of Paris casts of cicada emergence tunnels.

In the spring of the 17th year, the cicadas are four to six inches below the surface. Oftentimes you will see increased mole activity in yards due to the cicadas being at the same level as the hungry tunneling mammals. In late April and early May, following a heavy rain, the cicadas might build "chimneys" by packing balls of mud from inside their tunnel around the tunnel entrance. This extends their homes above the water-logged soil as the cicadas wait to emerge.

Nymphal chimneys: (a) front view; (b) section; (c) nymph awaiting metamorphosis; (d) nymph ready to transform; (e) opening. [After Riley, 1869]

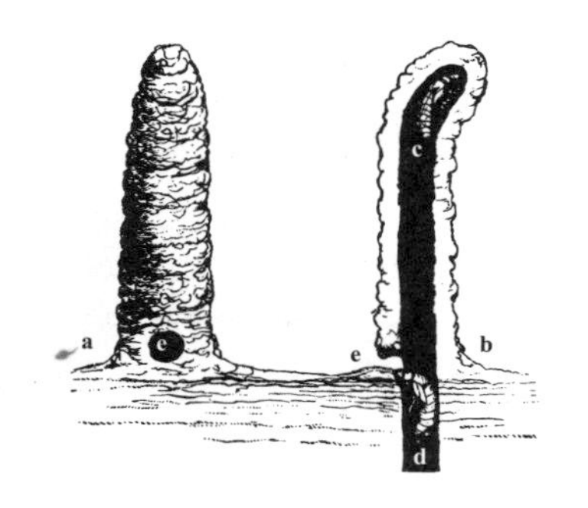


The mass emergence begins when the soil temperature reaches about 65°. Over the course of a few days, the periodical cicadas will emerge to start the process all over again.

## Suggested readings

For a detailed review of the biology see Williams and Simon (1995).

## Projects

You can study cicadas at anytime during their life cycle. Check the appendix to determine if they occur in your area. If they do, then you can do several different projects. If it is an emergence year in your area, start monitoring soil temperature. You can do this with a soil thermometer that can be purchased at a hardware store. Record the temperature under different trees and see at what temperature the cicadas emerge.

Look for the chimneys after a heavy spring rain. These will tell you where the cicadas will be emerging. If you find a chimney, then break off the top and see if the cicada is inside. Measure the height of several chimneys and calculate the average height.

After an emergence you can study the emergence tunnel by pouring plaster of Paris into the tunnel and letting it harden for 24 hours. Carefully excavate around the plaster to remove the cast of the cicada's tunnel.

After egg laying has been completed, take a twig that has a cicada oviposition scar on it and carefully cut away the bark to reveal the eggs. How many eggs were laid in the twig? Compare the number of eggs in each scar from different kinds of trees.

You can study the eggs up to four years after an emergence. Look for an oviposition scar on a tree and remove it for study. Carefully trim away the wood and you will find the eggs of the cicadas. Unhatched eggs will be solid in appearance whereas the hatched eggs will be transparent. Calculate the percentage of the eggs that successfully hatched. Do you see different hatch rates for different kinds of trees?

If you have periodical cicadas in your area, but it is not an emergence year, you can look for the cicadas underground. Ask people in the area if they remember cicadas being on specific trees. After getting permission to dig, carefully dig down 14 inches into the ground around the tree under the branches rather than right next to the trunk. The small cicadas can be found in tunnels along side the roots. Measure how deep the cicadas have tunneled. How many cicadas do you find in one square foot?

# Chapter 4. Species of Periodical Cicadas

People stopped thinking of periodical cicadas as a plague as naturalists started to study the insect scientifically. Pehr Kalm, a student of the Swedish scientist Carolus Linnaeus, described the life of the periodical cicada in 1756 and it was Linnaeus who described the first species of periodical cicada in 1759 naming it *Cicada septemdecim*. It was nearly a century later when a second smaller species was recognized and named *Cicada cassini*. The two species differed in their size, their song, and had slight differences in color. The smaller species was entirely black on its underside whereas the larger species had dark orange-red bands on its abdomen.

But how many species of periodical cicadas are there? Biologists define a species as all individuals that belong to potentially interbreeding populations. The periodical cicada species debate involves the different life cycles of 17- and 13-years. Every 221 years (17 X 13 = 221) in areas where 17- and 13-year cicadas overlap, they emerge together and the potential to interbreed occurs. Field studies have shown that during those years the periodical cicadas can interbreed. Does the life cycle difference qualify the 17-year and 13-year forms as different species because they can only interbreed every 221 years? Many think not. Charles Darwin was told about the periodical cicadas in a letter from his former college classmate, Benjamin Walsh, who was living in Illinois. Darwin replied to Walsh in 1868:

> ... I do not at all know what to think of your extraordinary case of the Cicadas. Professor Asa Gray and Dr. Hooker were staying here, and I told them of the facts. They thought that the 13- year and the 17-year forms ought not to be ranked as distinct species, unless other differences besides the period of development could be discovered. They thought the mere rarity of variability in such a point was not sufficient, and I think I concur with them....

However, Darwin's views did not make a significant impression on Walsh because he and his protégé, C.V. Riley, went on and described the 13-year cicadas as being different species from the 17-year cicadas. The larger 13-year form they named *Cicada tredecim* and the smaller 13-year form they named *Cicada tredecassini*. That brought the number of periodical cicada species to four.

In a revision of cicadas published in 1925, William T. Davis recognized that the periodical cicadas were different in appearance from the other cicadas that belonged to the genus *Cicada*. Because of the special life cycle and the public interest in these magical insects, he proposed a new genus for the periodical cicadas calling them *Magicicada*.

The number of periodical cicada species stood at four for nearly a century. But in 1962 Richard D. Alexander and Thomas E. Moore discovered a periodical cicada the same size as *Magicicada cassini* but with a different mating call and thin orange bands on the underside of the abdomen. They described the 17-year form as belonging to the species *Magicicada septendecula* and the 13-year form as the species *Magicicada tredecula*.

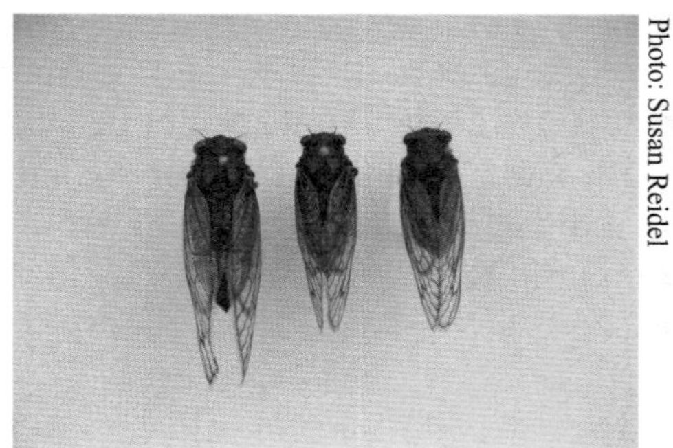

Photo: Susan Reidel

Top view of the three species of periodical cicadas. From left to right — *Magicicada septendecim, M. cassini, M. septendecula.*

Photo: Susan Reidel

Underside view of the three species of periodical cicadas. From left to right — *Magicicada septendecim, M. cassini, M. septendecula.*

But the separation of three distinct morphological forms with their different songs as six species based on their life cycle has been questioned. In 1998, the 17-year Brood IV and the 13-year Brood XIX emerged together in parts of Missouri and Iowa, something that only occurs every 221 years (17 times 13 equals 221). Careful surveys found the dual emergences were not synchronized and that the 17-year cicadas emerged at least ten days after the 13-year cicadas. This would support the view that there are indeed six species of periodical cicadas because the earlier emerging 13-year cicadas would have mated with each other by the time the 17-year cicadas emerged.

For Ohioans the problem is simple. Ohio does not have any 13-year cicada broods. Therefore, we only have three species of periodical cicadas living in our state regardless of which side of the species controversy you support. The Ohio species *are Magicicada septendecim, Magicicada cassini,* and *Magicicada septendecula.*

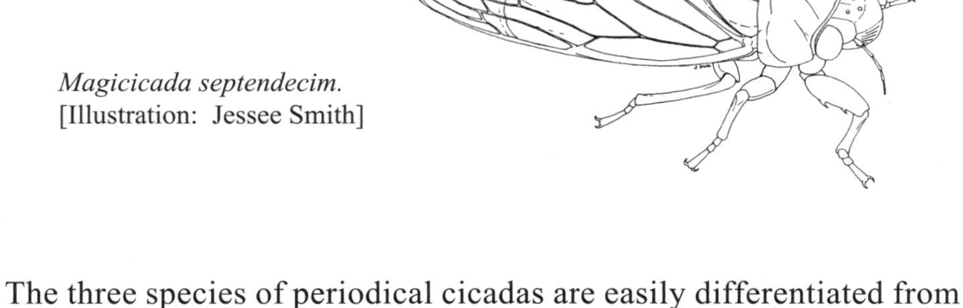

*Magicicada septendecim.*
[Illustration: Jessee Smith]

The three species of periodical cicadas are easily differentiated from the dog-day cicadas of summer. These annual cicadas are larger than their periodical cousins and have black eyes and a body that has a green and black pattern on the thorax. The dog-day cicadas usually are not heard until the end of June and during years when periodical cicadas also appear, they usually do not overlap. In 1987 when the periodical cicadas emerged in considerable numbers in Cincinnati, the last periodical cicada was heard on June 30 and the first dog-day cicada was recorded singing on July 1. The table on the next page can be used to help determine if the cicada you have found is a dog-day cicada or a periodical cicada.

Comparison of periodical cicadas and annual (dog-day) cicadas.

| Character | Periodical cicada | Dog-day cicada |
|---|---|---|
| Size | 1.0-1.5 inches | 1.75-2.0 inches |
| Body color | Black* | Green & black pattern |
| Eye color | Red | Black |
| Wing vein color | Orange-red | Green |
| Adult appearance | mid-May to mid June | late June through September |

* some with orange bands under the abdomen

Photographic comparison of a periodical cicada (left) and an annual (dog-day) cicada (right).

## Suggested readings

To study the species problem in detail see Alexander and Moore (1962), Williams and Simon (1995), Lloyd (1984), and Young (1958).

## Projects

If the periodical cicadas are emerging in your area, try and find the three different species. The cicadas are easily preserved in 70% ethanol or by pinning. If you are going to pin the insects use special insect pins available from natural history museum gift shops, parks, or hobby supply houses. Killing the cicadas can be accomplished by placing the insect in the ethanol if you plan to use ethanol as the preservative. If you are going to pin the cicadas, kill the adults by placing them in a jar or sandwich bag and putting it in the freezer overnight. The pin should be placed in the scutellum, the triangular structure between the wings.

# Chapter 5. Morphology of periodical cicadas

Periodical cicadas, being insects, have a body that is divided into three sections. These sections are the head, thorax, and abdomen. The most striking feature of the head is the large red eyes. Between and below the eyes is a swollen ribbed structure that bulges out from the head. This is the clypeus and inside are muscles which help the cicada drink. At the bottom of the clypeus are the needle-like mouthparts. The cicadas have sucking mouthparts that are enclosed inside a covering that looks like a tube protruding from the base of the clypeus and is oriented back between the front legs. If you use a needle you can carefully tease the needle-like mouthparts out of the covering. Between the eyes and the clypeus are the antennae. These are used by the cicada to detect information about its environment. The second section is the thorax and is where you will find the wings

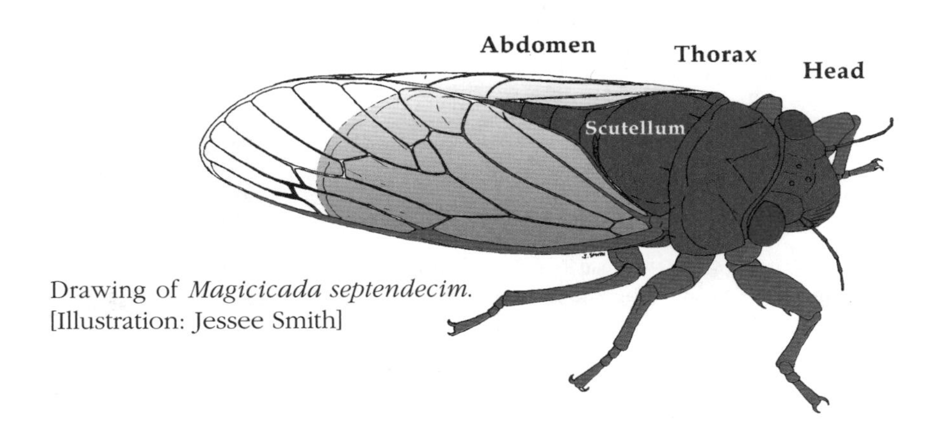

Drawing of *Magicicada septendecim.*
[Illustration: Jessee Smith]

and legs. The legs are orange and are typical walking legs as opposed to the digging forelegs that the nymphal cicada possesses. The wings are clear with orange veins.

If you look at the cicada from the top, you will see a large triangular plate between the wings. This is the scutellum and covers a thoracic cavity that is filled with muscles to power the cicada's flight.

Comparison of male and female cicadas. [Photo: Susan Reidel]

The abdomen is the last of the sections and houses the organs of digestion and reproduction and the male's sound producing organs. If you should dissect a male and compare it to a female you will find that the male's abdomen is essentially a hollow space which helps with sound production. Females, on the other hand, have an abdomen filled with eggs.

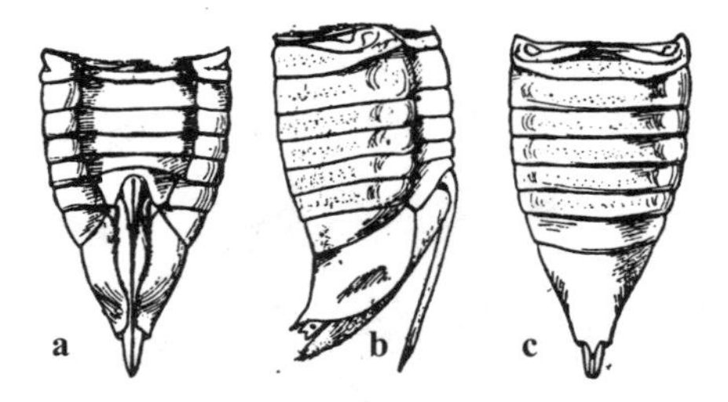


Above: Abdomen of female cicada showing ovipositor and attachments: a. ventral (bottom view); b. lateral (side view); c. dorsal (top view). [Illustration: Marlatt, 1907]

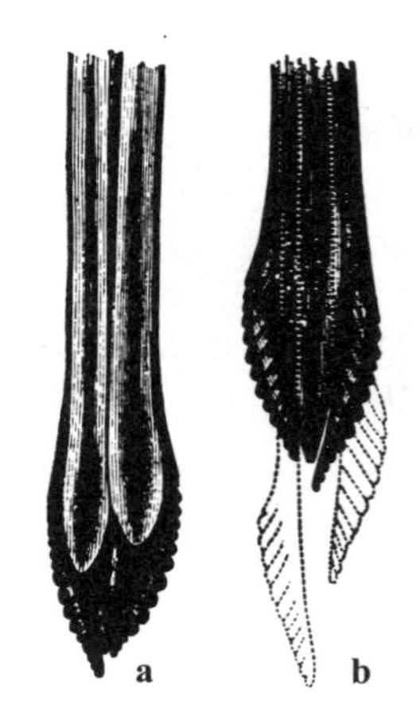


Right: Tip of ovipositor of female cicada (enlarged): a. dorsal (top) view; b. ventral (bottom view). Dotted portion shows alternating motion of the side pieces. [Illustration: Marlatt, 1907]

The sound producing structures, the tymbals, are easily seen on the male. Carefully lift up the wings and you will see two large white structures on each side of the top of the abdomen. If you look at them with a magnifying glass or microscope you will see these tymbals have ribs which help make the sounds.

## Suggested readings

To study the morphology of periodical cicadas in detail see Snodgrass (1921 and 1927).

**Project**

Catch a periodical or dog-day cicada and identify the parts. Label the diagram of the cicada as you find the structures on the insect. Make a photocopy of the diagram and color the insect.

# Chapter 6. Periodical Cicadas FAQs

## What is a periodical cicada?

Periodical cicadas are insects that belong to the order Hemiptera and the suborder Homoptera. They are relatives of leafhoppers, treehoppers, and aphids. Cicadas are large insects compared to most, and have red eyes and give loud mating calls. Their survival is linked to a mass appearance every 17, or in some areas, 13 years. This is part of their survival strategy. They appear in such great numbers that their predators cannot eat all of them. Indeed, their predators get tired of eating cicadas and those not consumed reproduce to continue the cycle.

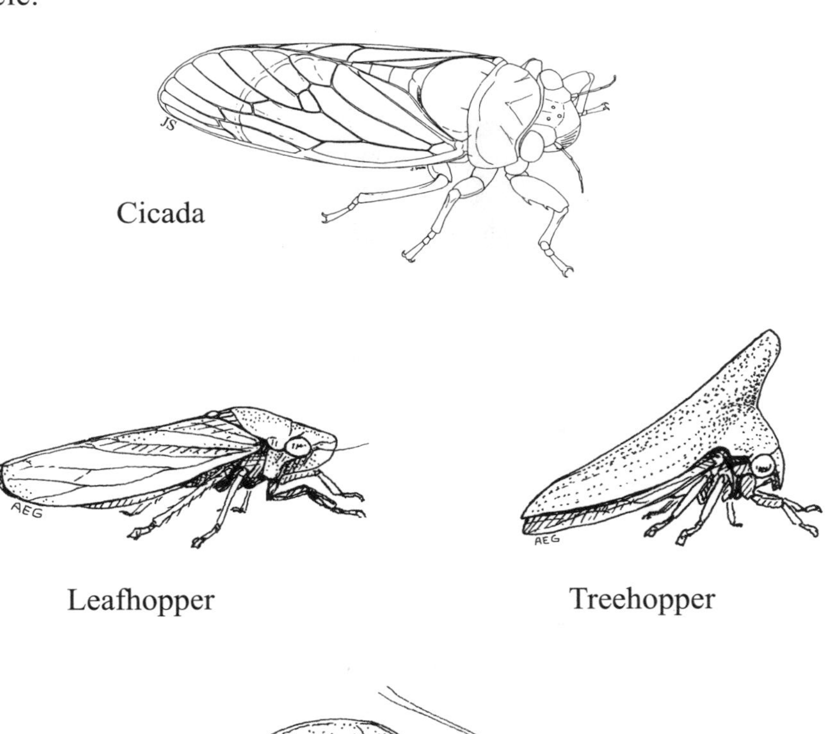


Cicada

Leafhopper

Treehopper

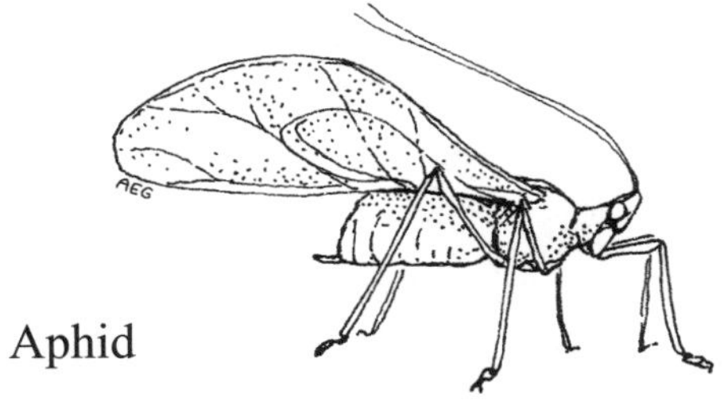


Aphid

## Why a 17-year life cycle?

The evolution of the 17-year life cycle is a mystery of the insect world. It is thought that the long life cycle evolved during the harsh times of the ice ages. One benefit of the long life cycle is that it makes it difficult for parasites to evolve a similar, synchronized life cycle.

## How can you tell the sexes apart?

It is easy to separate the males from the females. Look at the underside of the adult insect. Female cicadas have an egg-laying structure called an ovipositor and it is held within a slit-like opening at the tip of the female's abdomen. The males do not have this slit on the underside of the abdomen.

Female cicada on right. [Photo: Susan Reidel]

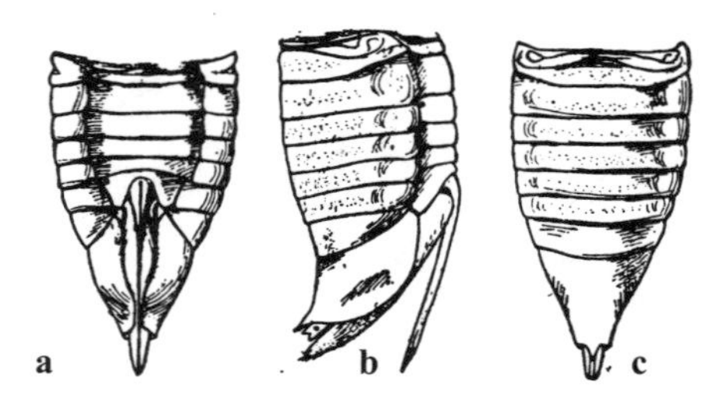


Abdomen of female cicada: a. ventral (bottom view); lateral (side view); and, dorsal (top view). [Ilustration: Marlatt, 1907]

## How do the cicadas make sound?

Only the male cicadas sing. They have sound producing structures called tymbals found on the anterior sides of the abdomen. These tymbals are ribbed with chitin, the material that makes up a cicada's exoskeleton. When these ribs are bent, it makes a sound which can be amplified by the abdomen which is mostly filled with air.

## Why do periodical cicadas sing?

The male cicadas sing to attract mates. They, however, do make other sounds. When threatened by a bird, a male cicada will make a warning call to startle its predator or to inform the attacker that it is a male. Male cicadas are mostly filled with air and do not make a good meal, whereas the female cicadas are filled with nutritious eggs.

## Can people eat cicadas?

Yes! The Iroquois used to harvest periodical cicadas and roasted them for food. In 1902 the Cincinnati *Enquirer* gave a recipe for a cicada pie which we have reprinted on the next page.

Iroquois woman roasting cicadas.

**Recipe for Periodical Cicada Pie**

Take 50 newly emerged white female cicadas and remove the wings, legs, and head. Chop up the cicadas into pieces and place in a bowl with stale bread that has been soaked in milk. Add sugar, rhubarb flavor, and cream to soften the ingredients. Put the mixture into a piecrust and cover with strips of piecrust placed in a cross pattern similar to that of an apple pie. Bake in an oven at 400° till crust is done.

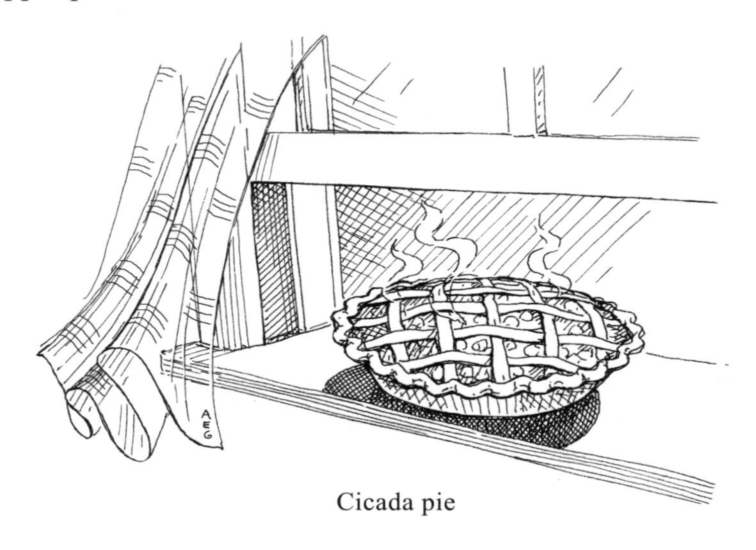

Cicada pie

People who have enjoyed this pie claim it tasted like partridge and that it "was good eating." Cincinnati *Enquirer*, June 6, 1902, page 4.

## What is flagging?

After the female cicada lays her eggs in the new growth of a tree, the twig sometimes breaks where the eggs were laid. The leaves on the broken twig die and turn brown. The drooping twigs are referred to as flags and the process is called flagging.

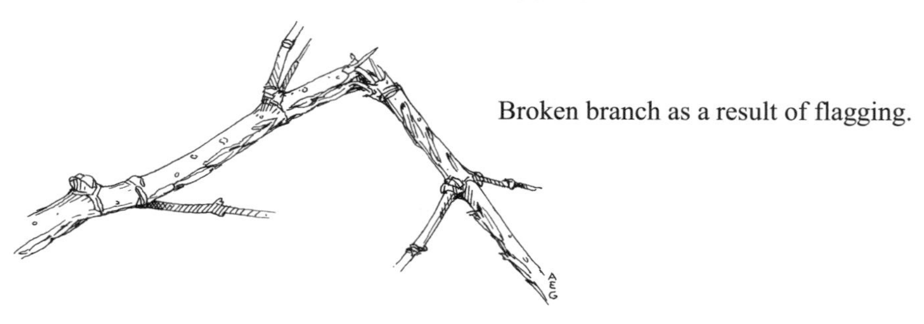

Broken branch as a result of flagging.

Tree in which flagging has occurred.

# Glossary

**abdomen:** the last of three major sections of the cicada's body; it contains the reproductive structures, and in adult females, is filled with eggs; the abdomen of an adult male is mostly filled with a large air space to help amplify the singing.

**brood:** all the cicadas that emerge in a particular year belong to the same brood; it is followed by a number that helps identify the year class of an emergence.

**chorusing center:** a large group of singing males in the same tree.

**clypeus:** the enlarged region between and below the eyes of cicadas; it is filled with muscles to help the cicada suck plant juices.

**emergence:** the synchronized appearance of adult periodical cicadas.

**flagging:** term used to describe the broken twigs and branches bearing dead leaves; caused by the female cicada's egg laying.

**head:** the first of three major sections of the cicada's body; it contains the eyes, antennae, and mouth parts.

**Hemiptera:** the order of insects that includes cicadas.

**Homoptera:** the suborder of Hemiptera that includes cicadas.

**locust:** a term incorrectly used for cicadas; true locusts are grasshoppers.

**molting:** the shedding of the nymphal skin by the developing adult cicada.

**nymph:** an immature cicada.

**ovipositor:** the egg laying structure of the cicada.

**scutellum:** a large triangular structure found between the wings and thorax.

**synchronized:** a term used to describe a simultaneous event, like an emergence.

**thorax:** the middle of three major sections of the cicada's body; it bears the legs and wings.

**trachea:** the air tubes inside the cicada that transport oxygen to its organs.

**tymbals:** the sound producing structures of male cicadas.

**USDA:** United States Department of Agriculture.

# APPENDICES

# APPENDIX A.
# Ohio's Emergence Records

The records for Ohio's periodical cicada emergence history were culled from reports received from the United States Department of Agriculture's Bureau of Entomology. This now extinct agency recorded three centuries of emergence records and preserved them on stenciled reports that were mailed to libraries around the country. Following World War II, the mapping efforts stopped except for isolated reports that appeared in various publications. In addition to the USDA records, I have included emergence data from several of the published papers listed in the bibliography.

The value of historical data is important in understanding how the broods evolve. Changes from the past distributions to current emergence patterns likely reflect declines in local populations due to agricultural practices and tree cutting. Other changes may reflect shifts in emergence years caused by 4-year accelerations and 1-year decelerations.

The records are listed by county, town, year, and brood. Broods are listed in Roman numerals except in those cases of accelerations and decelerations. They are given a prefix to indicate if they were an acceleration or a deceleration along with the brood that changed. To determine when the brood will next appear, look up the brood on the emergence table and scan across to a future year.

| County | Town | Year | Brood |
|---|---|---|---|
| Adams | | 1838 | XIV |
| Adams | | 1851 | X |
| Adams | | 1855 | XIV |
| Adams | | 1872 | XIV |
| Adams | | 1885 | X |
| Adams | | 1889 | XIV |
| Adams | | 1902 | X |
| Adams | | 1906 | XIV |
| Adams | | 1923 | XIV |
| Adams | | 1940 | XIV |
| Adams | Beaverpond | 1889 | XIV |
| Adams | Duncansville | 1940 | XIV |
| Adams | Harshasville | 1906 | XIV |
| Adams | Jefferson Twp. | 1940 | XIV |
| Adams | Lynx | 1995 | acc V |
| Adams | Manchester | 1923 | XIV |
| Adams | Manchester | 1957 | XIV |
| Adams | Mineral Springs | 1889 | XIV |
| Adams | Peebles | 1923 | XIV |
| Adams | Seaman | 1906 | XIV |
| Adams | Seaman | 1991 | XIV |
| Adams | Serpent Mound | 1940 | XIV |
| Adams | Serpent Mound | 1991 | XIV |
| Adams | Unity | 1991 | XIV |
| Adams | West Union | 1906 | XIV |
| Adams | West Union | 1923 | XIV |
| Adams | West Union | 1940 | XIV |
| Adams | West Union | 1987 | X |
| Adams | West Union | 1991 | XIV |
| Adams | Winchester | 1923 | XIV |
| Allen | | 1932 | acc X |
| Allen | | 1936 | X |
| Allen | Delphos | 1919 | X |
| Allen | Lima | 1902 | X |
| Ashland | | 1897 | V |
| Ashland | | 1914 | V |
| Ashland | | 1931 | V |
| Ashland | | 1965 | V |
| Athens | | 1897 | V |
| Athens | | 1914 | V |
| Athens | | 1931 | V |
| Athens | | 1965 | V |
| Athens | Athens | 1995 | acc V |
| Auglaize | | 1902 | X |
| Auglaize | | 1919 | X |
| Auglaize | | 1932 | acc X |
| Auglaize | | 1936 | X |
| Auglaize | Wapakoneta | 1885 | X |
| Auglaize | Wapakoneta | 1906 | XIV |
| Auglaize | Wapakoneta | 1987 | X |
| Belmont | | 1897 | V |
| Belmont | | 1914 | V |
| Belmont | | 1931 | V |
| Belmont | | 1965 | V |
| Brown | | 1804 | XIV |
| Brown | | 1821 | XIV |
| Brown | | 1838 | XIV |
| Brown | | 1936 | X |
| Brown | Aberdeen | 1957 | XIV |
| Brown | Franklin Twp. | 1906 | XIV |
| Brown | Georgetown | 1855 | XIV |
| Brown | Georgetown | 1872 | XIV |
| Brown | Georgetown | 1889 | XIV |
| Brown | Georgetown | 1906 | XIV |
| Brown | Georgetown | 1923 | XIV |
| Brown | Georgetown | 1991 | XIV |
| Brown | Hammersville | 1923 | XIV |
| Brown | Lewis Twp. | 1906 | XIV |
| Brown | Lewis Twp. | 1923 | XIV |
| Brown | Locustridge | 1906 | XIV |
| Brown | Mt. Orab | 1906 | XIV |
| Brown | Mt. Orab | 1923 | XIV |
| Brown | Mt. Orab | 1991 | XIV |
| Brown | Ripley | 1923 | XIV |
| Brown | Ripley | 1940 | XIV |
| Brown | Ripley | 1991 | XIV |
| Brown | Sardinia | 1906 | XIV |
| Brown | Sardinia | 1923 | XIV |
| Butler | | 1906 | XIV |
| Butler | | 1919 | X |
| Butler | | 1923 | XIV |
| Butler | | 1936 | X |
| Butler | | 1937 | dec X |
| Butler | | 1940 | XIV |
| Butler | Collinsville | 1902 | X |
| Butler | Excello | 1902 | X |
| Butler | Gano | 1872 | XIV |
| Butler | Hamilton | 1902 | X |
| Butler | Hamilton | 1991 | XIV |
| Butler | Indian Creek | 1937 | dec X |
| Butler | Overpeck | 1902 | X |
| Butler | Oxford | 1868 | X |
| Butler | Oxford | 1885 | X |
| Butler | Oxford | 1902 | X |
| Butler | Oxford | 1987 | X |
| Butler | Ross | 1902 | X |
| Butler | Ross | 1987 | X |
| Butler | West Chester | 1991 | XIV |
| Butler | Wood | 1834 | X |
| Butler | Wood | 1851 | X |
| Butler | Wood | 1868 | X |
| Butler | Wood | 1885 | X |
| Carroll | | 1849 | VIII |
| Carroll | | 1897 | V |
| Carroll | | 1900 | VIII |
| Carroll | | 1914 | V |
| Carroll | | 1917 | VIII |
| Carroll | | 1931 | V |
| Carroll | | 1934 | VIII |
| Carroll | | 1965 | V |
| Carroll | | 1968 | VIII |
| Carroll | Carrollton | 1898 | dec V |

| County | Town | Year | Brood |
|---|---|---|---|
| Champaign | | 1991 | XIV |
| Champaign | Mechanicsburg | 1919 | X |
| Champaign | Mechanicsburg | 1923 | XIV |
| Champaign | North Lewis | 1987 | X |
| Champaign | Rosewood | 1987 | X |
| Champaign | Urbana | 1885 | X |
| Champaign | Urbana | 1898 | acc X |
| Champaign | Urbana | 1902 | X |
| Champaign | Urbana | 1936 | X |
| Champaign | Woodstock | 1902 | X |
| Clark | | 1919 | X |
| Clark | | 1940 | XIV |
| Clark | Enon | 1902 | X |
| Clark | Forest Hills | 1987 | X |
| Clark | Lawrenceville | 1987 | X |
| Clark | Springfield | 1851 | X |
| Clark | Springfield | 1868 | X |
| Clark | Springfield | 1885 | X |
| Clark | Springfield | 1902 | X |
| Clark | Springfield | 1936 | X |
| Clark | Springfield | 1987 | X |
| Clark | St. Paris | 1987 | X |
| Clermont | | 1817 | X |
| Clermont | | 1834 | X |
| Clermont | | 1885 | X |
| Clermont | | 1936 | X |
| Clermont | | 1940 | XIV |
| Clermont | | 1957 | XIV |
| Clermont | | 1995 | ? |
| Clermont | Amelia | 1889 | XIV |
| Clermont | Amelia | 1906 | XIV |
| Clermont | Amelia | 1991 | XIV |
| Clermont | Batavia | 1923 | XIV |
| Clermont | Batavia | 1987 | X |
| Clermont | Batavia | 1991 | XIV |
| Clermont | Bethel | 1906 | XIV |
| Clermont | Bethel | 1991 | XIV |
| Clermont | E. Batavia Hts. | 1987 | X |
| Clermont | Felicity | 1991 | XIV |
| Clermont | Goshen | 1991 | XIV |
| Clermont | Hennings Mill | 1906 | XIV |
| Clermont | Loveland | 1889 | XIV |
| Clermont | Marathon | 1906 | XIV |
| Clermont | Miamiville | 1991 | XIV |
| Clermont | Milford | 1906 | XIV |
| Clermont | Milford | 1987 | X |
| Clermont | Milford | 1991 | XIV |
| Clermont | Moscow | 1988 | dec X |
| Clermont | New Richmond | 1923 | XIV |
| Clermont | New Richmond | 1991 | XIV |
| Clermont | Newtonsville | 1923 | XIV |
| Clermont | Nicholsville | 1906 | XIV |
| Clermont | Owensville | 1906 | XIV |
| Clermont | Perintown | 1991 | XIV |
| Clermont | Stonelick Twp. | 1923 | XIV |
| Clermont | Williamsburg | 1906 | XIV |
| Clinton | | 1868 | X |
| Clinton | | 1885 | X |
| Clinton | | 1902 | X |
| Clinton | | 1919 | X |
| Clinton | | 1936 | X |
| Clinton | | 1940 | XIV |
| Clinton | | 1991 | XIV |
| Clinton | Adams Twp. | 1923 | XIV |
| Clinton | Blanchester | 1906 | XIV |
| Clinton | Chester Twp. | 1923 | XIV |
| Clinton | Marion Twp. | 1855 | XIV |
| Clinton | Marion Twp. | 1872 | XIV |
| Clinton | Marion Twp. | 1906 | XIV |
| Clinton | Melvin | 1889 | XIV |
| Clinton | New Burlington | 1923 | XIV |
| Clinton | New Vienna | 1906 | XIV |
| Clinton | New Vienna | 1923 | XIV |
| Clinton | Palestine | 1851 | X |
| Clinton | Palestine | 1868 | X |
| Clinton | Palestine | 1885 | X |
| Clinton | Palestine | 1902 | X |
| Clinton | Union Twp. | 1923 | XIV |
| Clinton | Wayne Twp. | 1923 | XIV |
| Clinton | Wilmington | 1906 | XIV |
| Clinton | Wilmington | 1923 | XIV |
| Clinton | Wilmington | 1957 | XIV |
| Columbiana | | 1849 | VIII |
| Columbiana | | 1897 | V |
| Columbiana | | 1900 | VIII |
| Columbiana | | 1914 | V |
| Columbiana | | 1917 | VIII |
| Columbiana | | 1931 | V |
| Columbiana | | 1934 | VIII |
| Columbiana | | 1965 | V |
| Columbiana | | 1968 | VIII |
| Columbiana | Lisbon | 1898 | dec V |
| Columbiana | Lisbon | 1906 | XIV |
| Coshocton | | 1897 | V |
| Coshocton | | 1914 | V |
| Coshocton | | 1931 | V |
| Coshocton | | 1965 | V |
| Crawford | | 1919 | X |
| Crawford | Bucyrus | 1902 | X |
| Crawford | N. Robinson | 1936 | X |
| Cuyahoga | | 1897 | V |
| Cuyahoga | | 1914 | V |
| Cuyahoga | | 1931 | V |
| Cuyahoga | | 1965 | V |
| Cuyahoga | Brecksville | 1939 | ? |
| Darke | | 1919 | X |
| Darke | | 1936 | X |
| Darke | Greenville | 1834 | X |
| Darke | Greenville | 1851 | X |
| Darke | Greenville | 1868 | X |
| Darke | Greenville | 1885 | X |
| Darke | New Madison | 1987 | X |
| Darke | Palestine | 1851 | X |
| Darke | Palestine | 1868 | X |

| County | Town | Year | Brood |
|---|---|---|---|
| Darke | Palestine | 1885 | X |
| Darke | Palestine | 1902 | X |
| Defiance | Ayersville | 1987 | X |
| Delaware | | 1919 | X |
| Delaware | | 1931 | V |
| Delaware | | 1934 | VIII |
| Delaware | | 1940 | XIV |
| Delaware | Ashley | 1885 | X |
| Delaware | Ashley | 1902 | X |
| Delaware | Concord Twp. | 1987 | X |
| Delaware | Constantia | 1898 | acc X |
| Delaware | Delaware | 1885 | X |
| Delaware | Delaware | 1902 | X |
| Delaware | Delaware | 1906 | XIV |
| Delaware | Delaware | 1936 | X |
| Delaware | Harlem | 1987 | X |
| Delaware | Kilbourn | 1987 | X |
| Delaware | Liberty | 1987 | X |
| Delaware | Powell | 1902 | X |
| Delaware | Powell | 1987 | X |
| Delaware | Sunbury | 1885 | X |
| Delaware | Sunbury | 1902 | X |
| Erie | | 1897 | V |
| Erie | | 1914 | V |
| Erie | | 1965 | V |
| Fairfield | | 1897 | V |
| Fairfield | | 1914 | V |
| Fairfield | | 1919 | X |
| Fairfield | | 1931 | V |
| Fairfield | | 1965 | V |
| Fairfield | Amanda | 1902 | X |
| Fairfield | Bloom Twp. | 1936 | X |
| Fairfield | Canal Winchester | 1987 | X |
| Fairfield | Carroll | 1902 | X |
| Fairfield | Lancaster | 1868 | X |
| Fairfield | Lancaster | 1885 | X |
| Fairfield | Lancaster | 1902 | X |
| Fairfield | Lithopolis | 1868 | X |
| Fairfield | Lithopolis | 1885 | X |
| Fairfield | Lithopolis | 1902 | X |
| Fairfield | Lockville | 1902 | X |
| Fayette | | 1919 | X |
| Fayette | | 1987 | X |
| Fayette | | 1991 | XIV |
| Fayette | Concord Twp. | 1906 | XIV |
| Fayette | Green Twp. | 1906 | XIV |
| Fayette | Jeffersonville | 1906 | XIV |
| Fayette | Perry Twp. | 1906 | XIV |
| Fayette | Perry Twp. | 1923 | XIV |
| Fayette | Washington CH | 1872 | XIV |
| Fayette | Washington CH | 1906 | XIV |
| Fayette | Washington CH | 1936 | X |
| Fayette | Wayne Twp. | 1906 | XIV |
| Fayette | Wayne Twp. | 1923 | XIV |
| Franklin | | 1919 | X |
| Franklin | | 1931 | V |
| Franklin | | 1957 | XIV |
| Franklin | Briggsdale | 1902 | X |
| Franklin | Columbus | 1851 | X |
| Franklin | Columbus | 1885 | X |
| Franklin | Columbus | 1902 | X |
| Franklin | Columbus | 1936 | X |
| Franklin | Columbus | 1940 | XIV |
| Franklin | Columbus | 1987 | X |
| Franklin | Dublin | 1902 | X |
| Franklin | Dublin | 1987 | X |
| Franklin | Grandview Hts. | 1936 | X |
| Franklin | U. Arlington | 1936 | X |
| Gallia | | 1889 | XIV |
| Gallia | | 1897 | V |
| Gallia | | 1906 | XIV |
| Gallia | | 1914 | V |
| Gallia | | 1923 | XIV |
| Gallia | | 1931 | V |
| Gallia | | 1936 | X |
| Gallia | | 1940 | XIV |
| Gallia | | 1965 | V |
| Gallia | | 1991 | XIV |
| Gallia | | 1995 | acc V |
| Gallia | Bidwell | 1923 | XIV |
| Gallia | Clay Twp. | 1923 | XIV |
| Gallia | Cora | 1902 | X |
| Gallia | Gallia | 1906 | XIV |
| Gallia | Gallipolis | 1821 | XIV |
| Gallia | Gallipolis | 1923 | XIV |
| Gallia | Morgan Twp. | 1923 | XIV |
| Gallia | Northrup | 1923 | XIV |
| Gallia | Perry Twp. | 1923 | XIV |
| Gallia | Racoon Island | 1906 | XIV |
| Gallia | Rio Grande | 1957 | XIV |
| Gallia | Rodney | 1889 | XIV |
| Gallia | Thivener | 1923 | XIV |
| Gallia | Thurman | 1923 | XIV |
| Gallia | Vinton | 1923 | XIV |
| Gallia | Gallipolis | 1957 | XIV |
| Geauga | | 1897 | V |
| Geauga | | 1914 | V |
| Geauga | | 1931 | V |
| Geauga | | 1965 | V |
| Greene | | 1991 | XIV |
| Greene | Caesar Ck. Twp. | 1923 | XIV |
| Greene | Cedarville | 1919 | X |
| Greene | Clifton | 1936 | X |
| Greene | Clifton | 1940 | XIV |
| Greene | Jamestown | 1906 | XIV |
| Greene | Miami Twp. | 1902 | X |
| Greene | Spring Val. Twp. | 1906 | XIV |
| Greene | Spring Val. Twp. | 1923 | XIV |
| Greene | Xenia | 1885 | X |
| Greene | Xenia | 1906 | XIV |
| Greene | Xenia | 1923 | XIV |
| Greene | Yellow Spr. | 1834 | X |
| Greene | Yellow Spr. | 1851 | X |
| Greene | Yellow Spr. | 1868 | X |

| County | Town | Year | Brood |
|---|---|---|---|
| Greene | Yellow Spr. | 1889 | XIV |
| Greene | Yellow Spr. | 1987 | X |
| Guernsey | | 1897 | V |
| Guernsey | | 1914 | V |
| Guernsey | | 1931 | V |
| Guernsey | | 1936 | X |
| Guernsey | | 1965 | V |
| Hamilton | | 1898 | acc X |
| Hamilton | | 1919 | X |
| Hamilton | | 1940 | XIV |
| Hamilton | Addyston | 1987 | X |
| Hamilton | Amberley | 1991 | XIV |
| Hamilton | Anderson Twp. | 1923 | XIV |
| Hamilton | Avondale | 1902 | X |
| Hamilton | Blue Ash | 1987 | X |
| Hamilton | Blue Ash | 1991 | XIV |
| Hamilton | Bridgetown | 1987 | X |
| Hamilton | California | 1936 | X |
| Hamilton | Cheviot | 1987 | X |
| Hamilton | Cheviot | 1991 | XIV |
| Hamilton | Cincinnati | 1851 | X |
| Hamilton | Cincinnati | 1872 | XIV |
| Hamilton | Cincinnati | 1885 | X |
| Hamilton | Cincinnati | 1889 | XIV |
| Hamilton | Cincinnati | 1902 | X |
| Hamilton | Cincinnati | 1936 | X |
| Hamilton | Cincinnati | 1983 | acc X |
| Hamilton | Cincinnati | 1987 | X |
| Hamilton | Cincinnati | 1991 | XIV |
| Hamilton | Cleves | 1868 | X |
| Hamilton | Cleves | 1872 | XIV |
| Hamilton | Cleves | 1885 | X |
| Hamilton | Cleves | 1902 | X |
| Hamilton | Cleves | 1987 | X |
| Hamilton | Columbia Twp. | 1906 | XIV |
| Hamilton | Columbia Twp. | 1923 | XIV |
| Hamilton | Crescent | 1868 | X |
| Hamilton | Delhi | 1987 | X |
| Hamilton | Dent | 1987 | X |
| Hamilton | Dunlap | 1902 | X |
| Hamilton | Elizabethtown | 1987 | X |
| Hamilton | Fairfax | 1987 | X |
| Hamilton | Forest Park | 1987 | X |
| Hamilton | Goshen | 1987 | X |
| Hamilton | Greenhills | 1987 | X |
| Hamilton | Harrison | 1906 | XIV |
| Hamilton | Harrison | 1987 | X |
| Hamilton | Indian Hill | 1991 | XIV |
| Hamilton | Kenwood | 1991 | XIV |
| Hamilton | Loveland | 1906 | XIV |
| Hamilton | Loveland | 1987 | X |
| Hamilton | Loveland | 1995 | ? |
| Hamilton | Maderia | 1991 | XIV |
| Hamilton | Mariemont | 1987 | X |
| Hamilton | Mariemont | 1991 | XIV |
| Hamilton | Montgomery | 1987 | X |
| Hamilton | Montgomery | 1991 | XIV |
| Hamilton | Mt. Airy | 1936 | X |
| Hamilton | Mt. Airy | 1937 | dec X |
| Hamilton | Mt. Airy | 1987 | X |
| Hamilton | Mt. Healthy | 1987 | X |
| Hamilton | Mt. Washington | 1936 | X |
| Hamilton | Mt. St. Joseph | 1987 | X |
| Hamilton | N. College Hill | 1987 | X |
| Hamilton | Newton | 1991 | XIV |
| Hamilton | Plainville | 1906 | XIV |
| Hamilton | Sharonville | 1987 | X |
| Hamilton | Sharonville | 1991 | XIV |
| Hamilton | St. Bernard | 1987 | X |
| Hamilton | Sycamore Twp. | 1923 | XIV |
| Hamilton | Symmes Twp. | 1991 | XIV |
| Hamilton | Symmes Twp. | 1923 | XIV |
| Hamilton | Terrace Park | 1991 | XIV |
| Hamilton | Walnut Hills | 1868 | X |
| Hamilton | Walnut Hills | 1885 | X |
| Hamilton | Willowville | 1987 | X |
| Hamilton | Withrow Pres | 1988 | X |
| Hancock | | 1919 | X |
| Hancock | | 1936 | X |
| Hancock | McComb | 1902 | X |
| Hardin | | 1919 | X |
| Hardin | | 1936 | X |
| Harrison | | 1897 | V |
| Harrison | | 1914 | V |
| Harrison | | 1931 | V |
| Harrison | | 1965 | V |
| Highland | | 1872 | XIV |
| Highland | | 1940 | XIV |
| Highland | Belfast | 1923 | XIV |
| Highland | Concord Twp. | 1923 | XIV |
| Highland | Fort Hill | 1957 | XIV |
| Highland | Fort Hill | 1991 | XIV |
| Highland | Fort Hill | 1995 | acc V |
| Highland | Greenfield | 1906 | XIV |
| Highland | Greenfield | 1923 | XIV |
| Highland | Highland | 1923 | XIV |
| Highland | Hillsboro | 1906 | XIV |
| Highland | Hillsboro | 1923 | XIV |
| Highland | Hillsboro | 1923 | XIV |
| Highland | Hillsboro | 1991 | XIV |
| Highland | Jackson | 1923 | XIV |
| Highland | Leesburg | 1906 | XIV |
| Highland | Marshall | 1906 | XIV |
| Highland | Newmarket Twp. | 1923 | XIV |
| Highland | Paint Twp. | 1923 | XIV |
| Highland | Washington Twp. | 1923 | XIV |
| Highland | White Oak Twp. | 1923 | XIV |
| Hocking | | 1897 | V |
| Hocking | | 1914 | V |
| Hocking | | 1931 | V |
| Hocking | | 1956 | ? |
| Hocking | | 1961 | acc V |
| Hocking | | 1965 | V |
| Hocking | | 1995 | acc V |

| County | Town | Year | Brood |
|---|---|---|---|
| Hocking | Laurelville | 1885 | X |
| Holmes | | 1897 | V |
| Holmes | | 1914 | V |
| Holmes | | 1931 | V |
| Holmes | | 1965 | V |
| Huron | | 1897 | V |
| Huron | | 1902 | X |
| Huron | | 1914 | V |
| Huron | | 1919 | X |
| Huron | | 1931 | V |
| Huron | | 1965 | V |
| Jackson | | 1897 | V |
| Jackson | | 1914 | V |
| Jackson | | 1931 | V |
| Jackson | | 1961 | acc V |
| Jackson | | 1965 | V |
| Jackson | | 1991 | XIV |
| Jackson | Athalia | 1889 | XIV |
| Jackson | Coalton | 1902 | X |
| Jackson | Cove | 1906 | XIV |
| Jackson | Hanging Rock | 1906 | XIV |
| Jackson | Ironton | 1906 | XIV |
| Jackson | Ironton | 1923 | XIV |
| Jackson | Kitts Hill | 1923 | XIV |
| Jackson | Mason Twp. | 1906 | XIV |
| Jackson | North Kenova | 1906 | XIV |
| Jackson | Oak Hill | 1940 | XIV |
| Jackson | Proctorville | 1923 | XIV |
| Jackson | Rock Camp | 1906 | XIV |
| Jackson | Rock Camp | 1923 | XIV |
| Jackson | Sherritts | 1923 | XIV |
| Jackson | Wilgus | 1906 | XIV |
| Jefferson | | 1897 | V |
| Jefferson | | 1900 | VIII |
| Jefferson | | 1914 | V |
| Jefferson | | 1917 | VIII |
| Jefferson | | 1965 | V |
| Jefferson | | 1968 | VIII |
| Knox | | 1897 | V |
| Knox | | 1914 | V |
| Knox | | 1931 | V |
| Knox | | 1965 | V |
| Lake | | 1897 | V |
| Lake | | 1914 | V |
| Lake | | 1931 | V |
| Lake | | 1965 | V |
| Lawrence | | 1897 | V |
| Lawrence | | 1914 | V |
| Lawrence | | 1931 | V |
| Lawrence | | 1965 | V |
| Lawrence | Burlington | 1940 | XIV |
| Lawrence | Chesapeake | 1940 | XIV |
| Lawrence | Coal Grove | 1940 | XIV |
| Lawrence | Hanging Rock | 1940 | XIV |
| Lawrence | Ironton | 1940 | XIV |
| Lawrence | Ironton | 1991 | XIV |
| Lawrence | North Kenova | 1940 | XIV |
| Lawrence | Proctorville | 1940 | XIV |
| Lawrence | South Point | 1940 | XIV |
| Lawrence | Sybene | 1940 | XIV |
| Licking | | 1897 | V |
| Licking | | 1914 | V |
| Licking | | 1931 | V |
| Licking | | 1965 | V |
| Licking | Pataskala | 1919 | X |
| Logan | | 1919 | X |
| Logan | Belle Center | 1902 | X |
| Logan | Bellefontaine | 1936 | X |
| Logan | Bellefontaine | 1987 | X |
| Logan | East Liberty | 1987 | X |
| Logan | Huntsville | 1902 | X |
| Logan | Middleburg | 1987 | X |
| Logan | Quincy | 1902 | X |
| Logan | Rushylvania | 1987 | X |
| Logan | W. Mansfield | 1987 | X |
| Logan | Zanesfield | 1987 | X |
| Lorain | | 1897 | V |
| Lorain | | 1914 | V |
| Lorain | | 1931 | V |
| Lorain | | 1965 | V |
| Lucas | Toledo | 1851 | X |
| Lucas | Toledo | 1868 | X |
| Lucas | Toledo | 1885 | X |
| Lucas | Toledo | 1902 | X |
| Madison | | 1898 | acc X |
| Madison | | 1919 | X |
| Madison | | 1936 | X |
| Madison | Lilly Chapel | 1868 | X |
| Madison | Lilly Chapel | 1885 | X |
| Madison | Lilly Chapel | 1902 | X |
| Madison | Plain City | 1902 | X |
| Madison | W. Jefferson | 1902 | X |
| Madison | W. Jefferson | 1987 | X |
| Mahoning | | 1849 | VIII |
| Mahoning | | 1897 | V |
| Mahoning | | 1900 | VIII |
| Mahoning | | 1914 | V |
| Mahoning | | 1917 | VIII |
| Mahoning | | 1931 | V |
| Mahoning | | 1968 | VIII |
| Mahoning | Berlin Center | 1898 | dec V |
| Mahoning | Youngstown | 1934 | VIII |
| Marion | | 1919 | X |
| Marion | | 1936 | X |
| Marion | Marion Twp. | 1885 | X |
| Marion | Marion Twp. | 1902 | X |
| Marion | Morral | 1902 | X |
| Marion | Pleasant Twp. | 1885 | X |
| Marion | Pleasant Twp. | 1902 | X |
| Marion | Prospect | 1902 | X |
| Medina | | 1897 | V |
| Medina | | 1914 | V |
| Medina | | 1931 | V |
| Medina | | 1965 | V |

| County | Town | Year | Brood |
|---|---|---|---|
| Meigs | | 1897 | V |
| Meigs | | 1906 | XIV |
| Meigs | | 1914 | V |
| Meigs | | 1931 | V |
| Meigs | | 1940 | XIV |
| Meigs | | 1965 | V |
| Meigs | Alfred | 1923 | XIV |
| Meigs | Pomeroy | 1923 | XIV |
| Meigs | Rutland | 1872 | XIV |
| Mercer | | 1919 | X |
| Mercer | | 1936 | X |
| Mercer | Rockford | 1902 | X |
| Miami | | 1936 | X |
| Miami | Piqua | 1851 | X |
| Miami | Piqua | 1902 | X |
| Miami | Piqua | 1919 | X |
| Miami | Troy | 1885 | X |
| Monroe | | 1897 | V |
| Monroe | | 1914 | V |
| Monroe | | 1931 | V |
| Monroe | | 1965 | V |
| Montgomery | | 1898 | acc X |
| Montgomery | | 1940 | XIV |
| Montgomery | Centerville | 1902 | X |
| Montgomery | Centerville | 1919 | X |
| Montgomery | Centerville | 1987 | X |
| Montgomery | Dayton | 1885 | X |
| Montgomery | Dayton | 1902 | X |
| Montgomery | Dayton | 1936 | X |
| Montgomery | Dayton | 1987 | X |
| Montgomery | Germantown | 1902 | X |
| Montgomery | Miamisburg | 1868 | X |
| Montgomery | Miamisburg | 1885 | X |
| Montgomery | Miamisburg | 1987 | X |
| Montgomery | Troutwood | 1987 | X |
| Montgomery | Vandalia | 1987 | X |
| Montgomery | W. Carrollton | 1987 | X |
| Morgan | | 1897 | V |
| Morgan | | 1914 | V |
| Morgan | | 1931 | V |
| Morgan | | 1965 | V |
| Morrow | | 1897 | V |
| Morrow | | 1902 | X |
| Morrow | | 1914 | V |
| Morrow | | 1919 | X |
| Morrow | | 1965 | V |
| Morrow | Cardington | 1987 | X |
| Morrow | Marengo | 1987 | X |
| Muskingum | | 1897 | V |
| Muskingum | | 1914 | V |
| Muskingum | | 1931 | V |
| Muskingum | | 1965 | V |
| Noble | | 1897 | V |
| Noble | | 1914 | V |
| Noble | | 1931 | V |
| Noble | | 1965 | V |
| Paulding | | 1932 | acc X |
| Paulding | NE corner | 1987 | X |
| Perry | | 1897 | V |
| Perry | | 1914 | V |
| Perry | | 1931 | V |
| Perry | | 1965 | V |
| Pickaway | | 1897 | V |
| Pickaway | | 1914 | V |
| Pickaway | | 1919 | X |
| Pickaway | | 1940 | XIV |
| Pickaway | | 1957 | XIV |
| Pickaway | | 1957 | XIV |
| Pickaway | | 1965 | V |
| Pickaway | Ashville | 1902 | X |
| Pickaway | Darbyville | 1898 | acc X |
| Pickaway | Darbyville | 1902 | X |
| Pickaway | Derby | 1936 | X |
| Pickaway | Nebraska | 1885 | X |
| Pickaway | Orient | 1987 | X |
| Pickaway | Pherson | 1885 | X |
| Pickaway | S. Bloomfield | 1936 | X |
| Pickaway | Washington | 1956 | ? |
| Pickaway | Williamsport | 1936 | X |
| Pike | | 1897 | V |
| Pike | | 1914 | V |
| Pike | | 1923 | XIV |
| Pike | | 1931 | V |
| Pike | | 1965 | V |
| Pike | | 1991 | XIV |
| Pike | Camp | 1906 | XIV |
| Pike | Idaho | 1906 | XIV |
| Pike | Morgantown | 1940 | XIV |
| Pike | Piketon | 1940 | XIV |
| Pike | Sargents | 1940 | XIV |
| Pike | Seal Twp. | 1906 | XIV |
| Pike | Wakefield | 1940 | XIV |
| Pike | Waverely | 1889 | XIV |
| Pike | Lake White | 1940 | XIV |
| Portage | | 1897 | V |
| Portage | | 1900 | VIII |
| Portage | | 1914 | V |
| Portage | | 1917 | VIII |
| Portage | | 1931 | V |
| Portage | | 1965 | V |
| Portage | | 1968 | VIII |
| Preble | | 1919 | X |
| Preble | | 1936 | X |
| Preble | | 1987 | X |
| Preble | Israel Twp. | 1906 | XIV |
| Preble | Lewisburg | 1834 | X |
| Preble | Lewisburg | 1885 | X |
| Preble | W. Alexandria | 1902 | X |
| Putnam | Columbus Grove | 1902 | X |
| Richland | | 1897 | V |
| Richland | | 1914 | V |
| Richland | | 1931 | V |
| Richland | | 1932 | dec V |
| Richland | | 1965 | V |

| County | Town | Year | Brood |
|---|---|---|---|
| Ross | | 1897 | V |
| Ross | | 1914 | V |
| Ross | | 1931 | V |
| Ross | | 1965 | V |
| Ross | | 1991 | XIV |
| Ross | Bainbridge | 1906 | XIV |
| Ross | Bourneville | 1923 | XIV |
| Ross | Buckskin Twp. | 1923 | XIV |
| Ross | Chillicothe | 1906 | XIV |
| Ross | Chillicothe | 1923 | XIV |
| Ross | Chillicothe | 1940 | XIV |
| Ross | Chillicothe | 1957 | XIV |
| Ross | Clarksburg | 1957 | XIV |
| Ross | Concord Twp. | 1923 | XIV |
| Ross | Frankfort | 1872 | XIV |
| Ross | Frankfort | 1923 | XIV |
| Ross | Lyndon | 1906 | XIV |
| Ross | Roxabell | 1906 | XIV |
| Ross | Roxabell | 1923 | XIV |
| Ross | Twin Twp. | 1923 | XIV |
| Sandusky | Fremont | 1885 | X |
| Scioto | | 1897 | V |
| Scioto | | 1914 | V |
| Scioto | | 1931 | V |
| Scioto | | 1965 | V |
| Scioto | Franklin Furn. | 1906 | XIV |
| Scioto | Franklin Furn. | 1940 | XIV |
| Scioto | Haverhill | 1940 | XIV |
| Scioto | Lucasville | 1906 | XIV |
| Scioto | Lucasville | 1923 | XIV |
| Scioto | Lucasville | 1940 | XIV |
| Scioto | New Boston | 1940 | XIV |
| Scioto | Nile Twp. | 1940 | XIV |
| Scioto | Otway | 1906 | XIV |
| Scioto | Otway | 1923 | XIV |
| Scioto | Portsmouth | 1906 | XIV |
| Scioto | Portsmouth | 1923 | XIV |
| Scioto | Portsmouth | 1940 | XIV |
| Scioto | Portsmouth | 1957 | XIV |
| Scioto | Portsmouth | 1991 | XIV |
| Scioto | Rarden | 1923 | XIV |
| Scioto | Sciotoville | 1906 | XIV |
| Scioto | Wheelersburg | 1906 | XIV |
| Scioto | Wheelersburg | 1923 | XIV |
| Scioto | Wheelersburg | 1940 | XIV |
| Seneca | | 1834 | X |
| Seneca | | 1851 | X |
| Seneca | | 1868 | X |
| Seneca | | 1897 | V |
| Seneca | | 1914 | V |
| Seneca | | 1919 | X |
| Seneca | | 1965 | V |
| Seneca | Flat Rock | 1902 | X |
| Shelby | | 1919 | X |
| Shelby | Dawson | 1898 | acc X |
| Shelby | Dawson | 1902 | X |
| Shelby | Fort Loramie | 1987 | X |
| Shelby | Kettlersville | 1936 | X |
| Shelby | Newport | 1987 | X |
| Shelby | Pemberton | 1902 | X |
| Shelby | Russia | 1987 | X |
| Shelby | Sidney | 1885 | X |
| Shelby | Sidney | 1902 | X |
| Shelby | Sidney | 1987 | X |
| Stark | | 1897 | V |
| Stark | | 1900 | VIII |
| Stark | | 1914 | V |
| Stark | | 1917 | VIII |
| Stark | | 1931 | V |
| Stark | | 1934 | VIII |
| Stark | | 1965 | V |
| Stark | Canton | 1939 | ? |
| Summit | | 1897 | V |
| Summit | | 1914 | V |
| Summit | | 1931 | V |
| Summit | | 1965 | V |
| Trumbull | | 1897 | V |
| Trumbull | | 1900 | VIII |
| Trumbull | | 1914 | V |
| Trumbull | | 1917 | VIII |
| Trumbull | | 1949 | acc X |
| Trumbull | | 1968 | VIII |
| Tuscarawas | | 1897 | V |
| Tuscarawas | | 1914 | V |
| Tuscarawas | | 1917 | VIII |
| Tuscarawas | | 1931 | V |
| Tuscarawas | | 1965 | V |
| Union | | 1919 | X |
| Union | | 1987 | X |
| Union | Milford Center | 1898 | acc X |
| Union | Milford Center | 1936 | X |
| Union | Richwood | 1902 | X |
| Van Wert | | 1919 | X |
| Van Wert | | 1932 | acc X |
| Van Wert | | 1936 | X |
| Van Wert | Willshire | 1902 | X |
| Vinton | | 1853 | ? |
| Vinton | | 1855 | XIV |
| Vinton | | 1897 | V |
| Vinton | | 1906 | XIV |
| Vinton | | 1914 | V |
| Vinton | | 1931 | V |
| Vinton | | 1965 | V |
| Vinton | McArthur | 1906 | XIV |
| Vinton | McArthur | 1923 | XIV |
| Warren | | 1906 | XIV |
| Warren | | 1919 | X |
| Warren | | 1936 | X |
| Warren | | 1940 | XIV |
| Warren | Clarksville | 1957 | XIV |
| Warren | Cozaddale | 1889 | XIV |
| Warren | Fort Ancient | 1940 | XIV |
| Warren | Fort Ancient | 1987 | X |
| Warren | Fort Ancient | 1991 | XIV |

| County | Town | Year | Brood |
|---|---|---|---|
| Warren | Foster | 1991 | XIV |
| Warren | Franklin | 1902 | X |
| Warren | Franklin | 1923 | XIV |
| Warren | Harlan Twp. | 1906 | XIV |
| Warren | Harlan Twp. | 1923 | XIV |
| Warren | Harveysburg | 1906 | XIV |
| Warren | Lebanon | 1885 | X |
| Warren | Lebanon | 1902 | X |
| Warren | Lebanon | 1906 | XIV |
| Warren | Lebanon | 1923 | XIV |
| Warren | Maineville | 1991 | XIV |
| Warren | Middletown Jct. | 1923 | XIV |
| Warren | Oregonia | 1923 | XIV |
| Warren | Oregonia | 1991 | XIV |
| Warren | Pleasant Plain | 1923 | XIV |
| Warren | Turtle Ck. Twp. | 1923 | XIV |
| Warren | Waynesville | 1838 | XIV |
| Warren | Waynesville | 1855 | XIV |
| Warren | Waynesville | 1872 | XIV |
| Warren | Waynesville | 1923 | XIV |
| Washington | | 1897 | V |
| Washington | | 1914 | V |
| Washington | | 1931 | V |
| Washington | | 1961 | acc V |
| Washington | | 1965 | V |
| Washington | | 1991 | XIV |
| Washington | Bartlett | 1923 | XIV |
| Washington | Fleming | 1923 | XIV |
| Washington | Little Hocking | 1906 | XIV |
| Washington | Marietta | 1995 | acc V |
| Wayne | | 1897 | V |
| Wayne | | 1914 | V |
| Wayne | | 1931 | V |
| Wayne | | 1965 | V |
| Williams | | 1987 | X |
| Wyandot | | 1919 | X |
| Wyandot | | 1936 | X |
| Wyandot | | 1987 | X |
| Wyandot | Carey | 1902 | X |
| Wyandot | Lovell | 1885 | X |
| Wyandot | U. Sandusky | 1868 | X |
| Wyandot | U. Sandusky | 1885 | X |
| Wyandot | U. Sandusky | 1902 | X |

# APPENDIX B.
# Emergence Table

Past and future emergence years for 17-year and 13-year periodical cicada broods.

## 17-year broods

| I | II | III | IV | V | VI | VII | VIII | IX |
|---|---|---|---|---|---|---|---|---|
| 1621 | 1622 | 1623 | 1624 | 1625 | 1626 | 1627 | 1628 | 1629 |
| 1638 | 1639 | 1640 | 1641 | 1642 | 1643 | 1644 | 1645 | 1646 |
| 1655 | 1656 | 1657 | 1658 | 1659 | 1660 | 1661 | 1662 | 1663 |
| 1672 | 1673 | 1674 | 1675 | 1676 | 1677 | 1678 | 1679 | 1680 |
| 1689 | 1690 | 1691 | 1692 | 1693 | 1694 | 1695 | 1696 | 1697 |
| 1706 | 1707 | 1708 | 1709 | 1710 | 1711 | 1712 | 1713 | 1714 |
| 1723 | 1724 | 1725 | 1726 | 1727 | 1728 | 1729 | 1730 | 1731 |
| 1740 | 1741 | 1742 | 1743 | 1744 | 1745 | 1746 | 1747 | 1748 |
| 1757 | 1758 | 1759 | 1760 | 1761 | 1762 | 1763 | 1764 | 1765 |
| 1774 | 1775 | 1776 | 1777 | 1778 | 1779 | 1780 | 1781 | 1782 |
| 1791 | 1792 | 1793 | 1794 | 1795 | 1796 | 1797 | 1798 | 1799 |
| 1808 | 1809 | 1810 | 1811 | 1812 | 1813 | 1814 | 1815 | 1816 |
| 1825 | 1826 | 1827 | 1828 | 1829 | 1830 | 1831 | 1832 | 1833 |
| 1842 | 1843 | 1844 | 1845 | 1846 | 1847 | 1848 | 1849 | 1850 |
| 1859 | 1860 | 1861 | 1862 | 1863 | 1864 | 1865 | 1866 | 1867 |
| 1876 | 1877 | 1878 | 1879 | 1880 | 1881 | 1882 | 1883 | 1884 |
| 1893 | 1894 | 1895 | 1896 | 1897 | 1898 | 1899 | 1900 | 1901 |
| 1910 | 1911 | 1912 | 1913 | 1914 | 1915 | 1916 | 1917 | 1918 |
| 1927 | 1928 | 1929 | 1930 | 1931 | 1932 | 1933 | 1934 | 1935 |
| 1944 | 1945 | 1946 | 1947 | 1948 | 1949 | 1950 | 1951 | 1952 |
| 1961 | 1962 | 1963 | 1964 | 1965 | 1966 | 1967 | 1968 | 1969 |
| 1978 | 1979 | 1980 | 1981 | 1982 | 1983 | 1984 | 1985 | 1986 |
| 1995 | 1996 | 1997 | 1998 | 1999 | 2000 | 2001 | 2002 | 2003 |
| 2012 | 2013 | 2014 | 2015 | 2016 | 2017 | 2018 | 2019 | 2020 |
| 2029 | 2030 | 2031 | 2032 | 2033 | 2034 | 2035 | 2036 | 2037 |
| 2046 | 2047 | 2048 | 2049 | 2050 | 2051 | 2052 | 2053 | 2054 |

Past and future emergence years for 17-year and 13-year periodical cicada broods, continued.

| 17-year broods | | | | 13-year broods | | |
|---|---|---|---|---|---|---|
| X | XI | XIII | XIV | XIX | XXII | XXIII |
| 1630 | 1631 | 1633 | 1634 | 1686 | 1689 | 1690 |
| 1647 | 1648 | 1650 | 1651 | 1699 | 1702 | 1703 |
| 1664 | 1665 | 1667 | 1668 | 1712 | 1715 | 1716 |
| 1681 | 1682 | 1684 | 1685 | 1725 | 1728 | 1729 |
| 1698 | 1699 | 1701 | 1702 | 1738 | 1741 | 1742 |
| 1715 | 1716 | 1718 | 1719 | 1751 | 1754 | 1755 |
| 1732 | 1733 | 1735 | 1736 | 1764 | 1767 | 1768 |
| 1749 | 1750 | 1752 | 1753 | 1777 | 1780 | 1781 |
| 1766 | 1767 | 1769 | 1770 | 1790 | 1793 | 1794 |
| 1783 | 1784 | 1786 | 1787 | 1803 | 1806 | 1807 |
| 1800 | 1801 | 1803 | 1804 | 1816 | 1819 | 1820 |
| 1817 | 1818 | 1820 | 1821 | 1829 | 1832 | 1833 |
| 1834 | 1835 | 1837 | 1838 | 1842 | 1845 | 1846 |
| 1851 | 1852 | 1854 | 1855 | 1855 | 1858 | 1859 |
| 1868 | 1869 | 1871 | 1872 | 1868 | 1871 | 1872 |
| 1885 | 1886 | 1888 | 1889 | 1881 | 1884 | 1885 |
| 1902 | 1903 | 1905 | 1906 | 1894 | 1897 | 1898 |
| 1919 | 1920 | 1922 | 1923 | 1907 | 1910 | 1911 |
| 1936 | 1937 | 1939 | 1940 | 1920 | 1923 | 1924 |
| 1953 | 1954 | 1956 | 1957 | 1933 | 1936 | 1937 |
| 1970 | 1971 | 1973 | 1974 | 1946 | 1949 | 1950 |
| 1987 | 1988 | 1990 | 1991 | 1959 | 1962 | 1963 |
| 2004 | 2005 | 2007 | 2008 | 1972 | 1975 | 1976 |
| 2021 | 2022 | 2024 | 2025 | 1985 | 1988 | 1989 |
| 2038 | 2039 | 2041 | 2042 | 1998 | 2001 | 2002 |
| 2055 | 2056 | 2058 | 2059 | 2011 | 2014 | 2015 |

# APPENDIX C.
# Distribution Maps for U.S. Broods

Distribution maps for United States' cicada broods are presented in this appendix. The map below shows the distribution of all 17-year (blue dots) and 13-year (red dots) cicada broods.

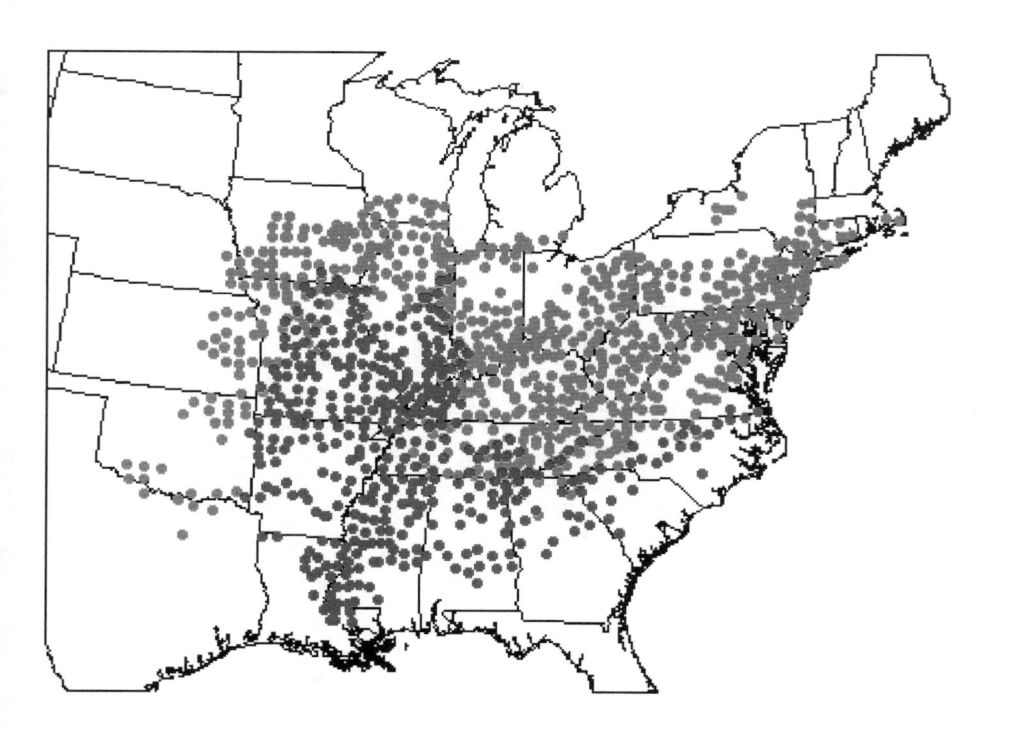

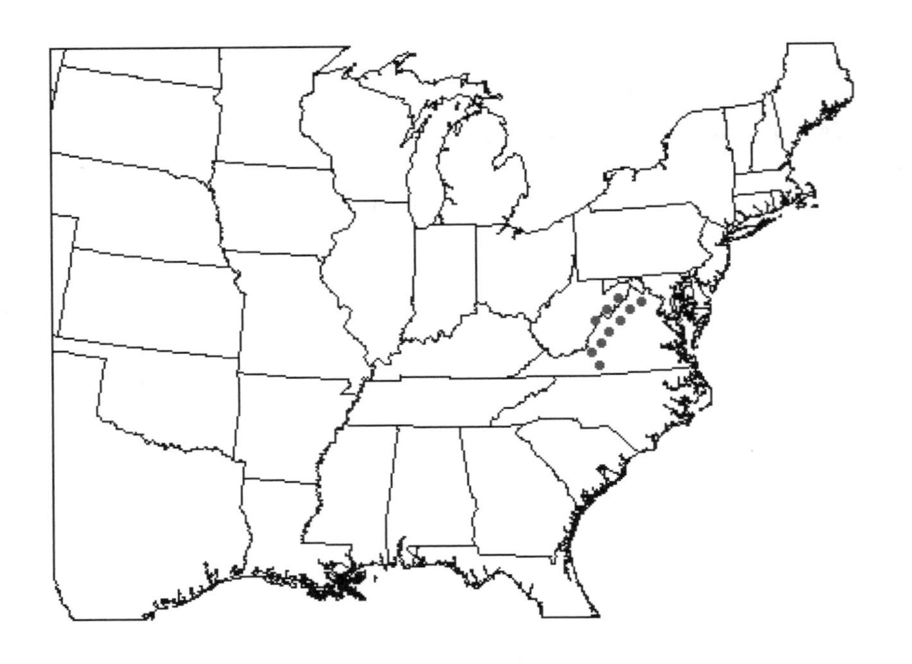

17-year cicada:  Brood I distribution.

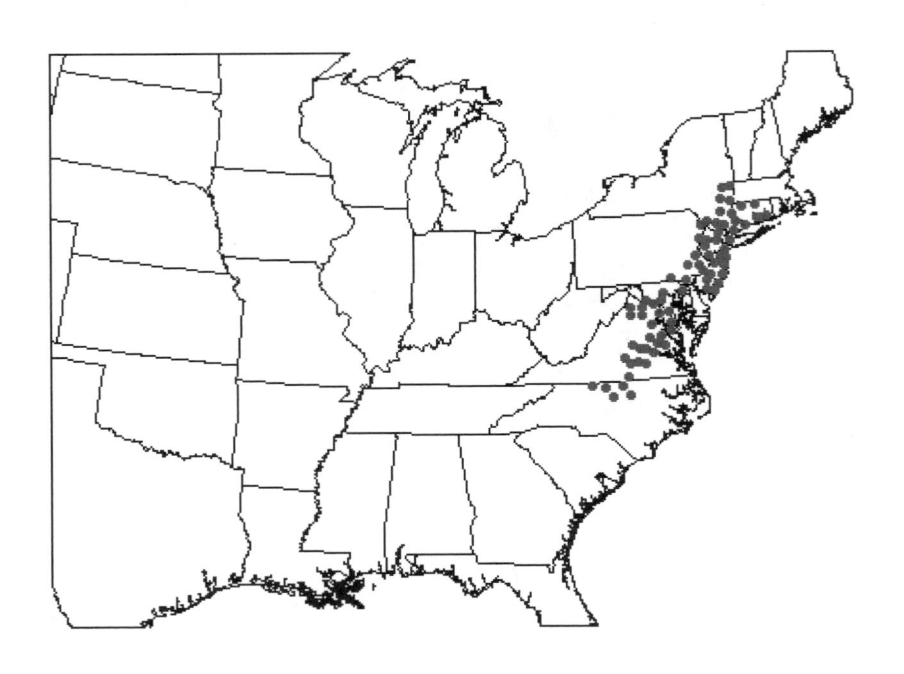

17-year cicada:  Brood II distribution.

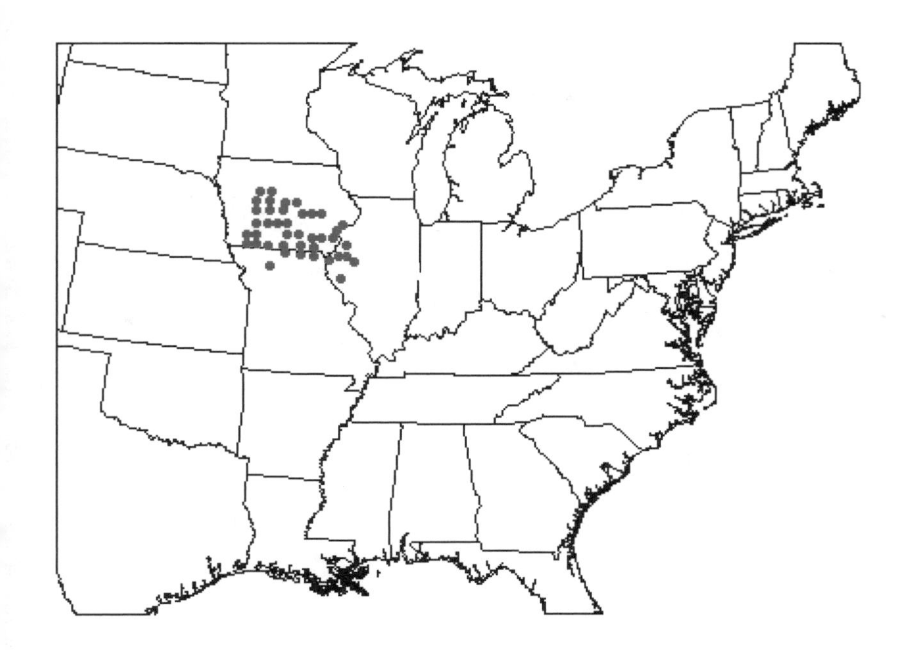

17-year cicada:  Brood III distribution.

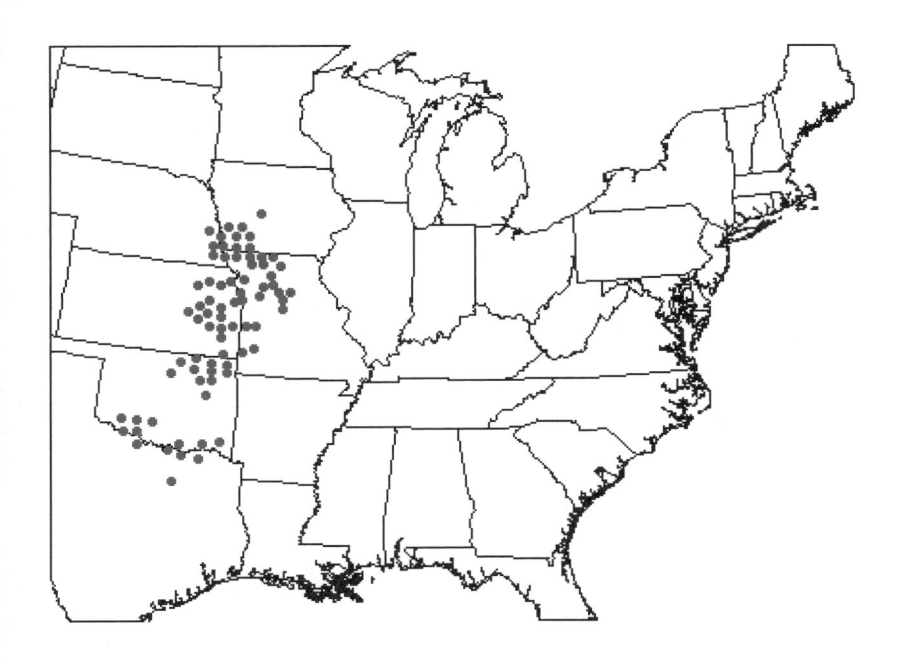

17-year cicada:  Brood IV distribution.

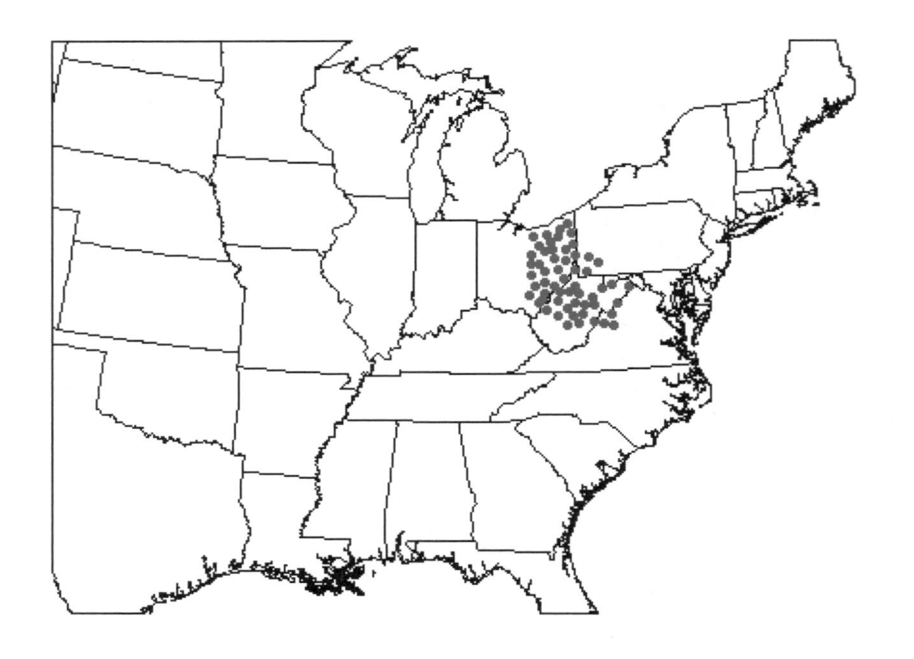

17-year cicada: Brood V distribution.

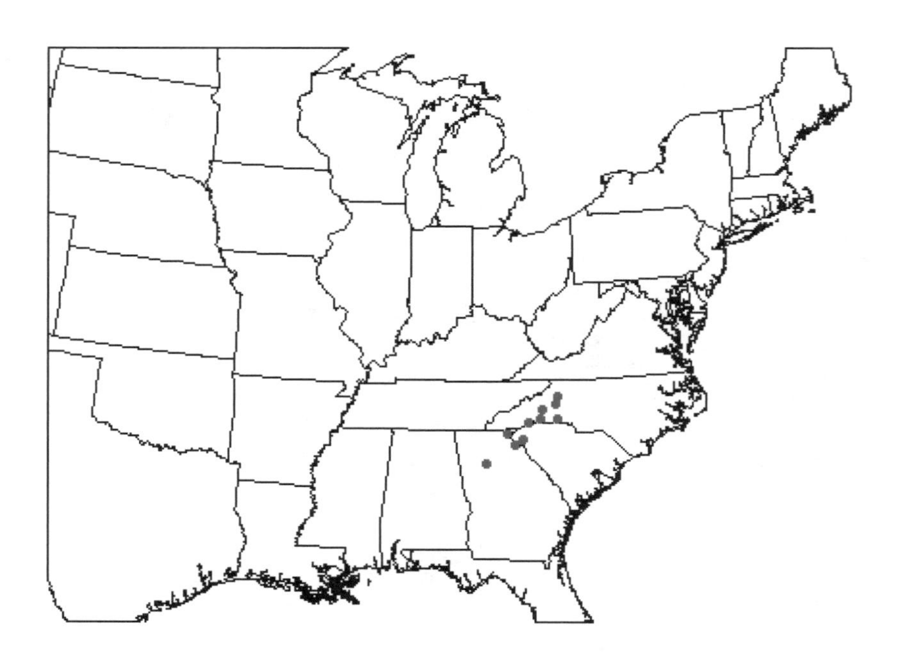

17-year cicada: Brood VI distribution.

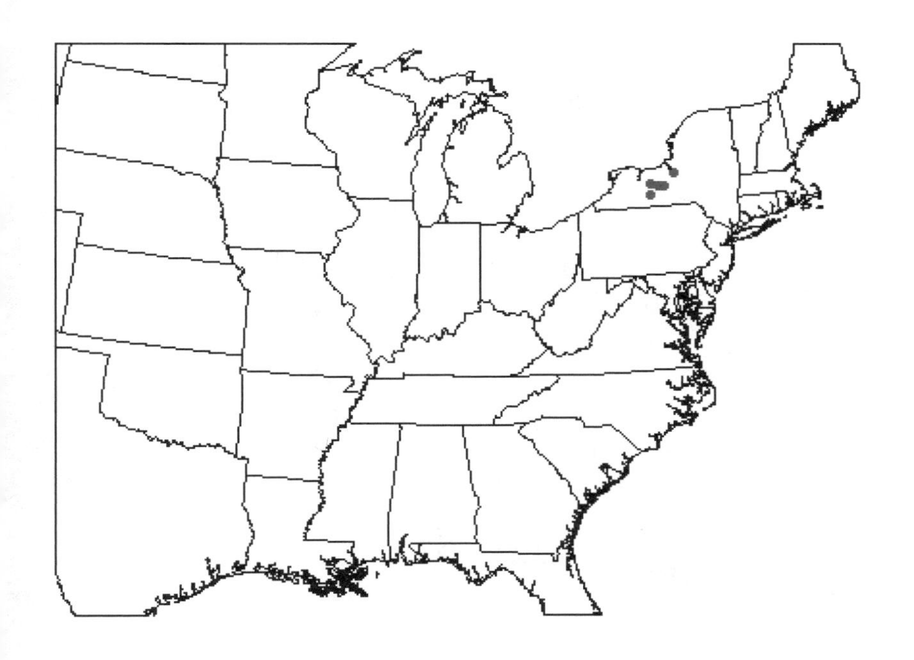

17-year cicada:  Brood VII distribution.

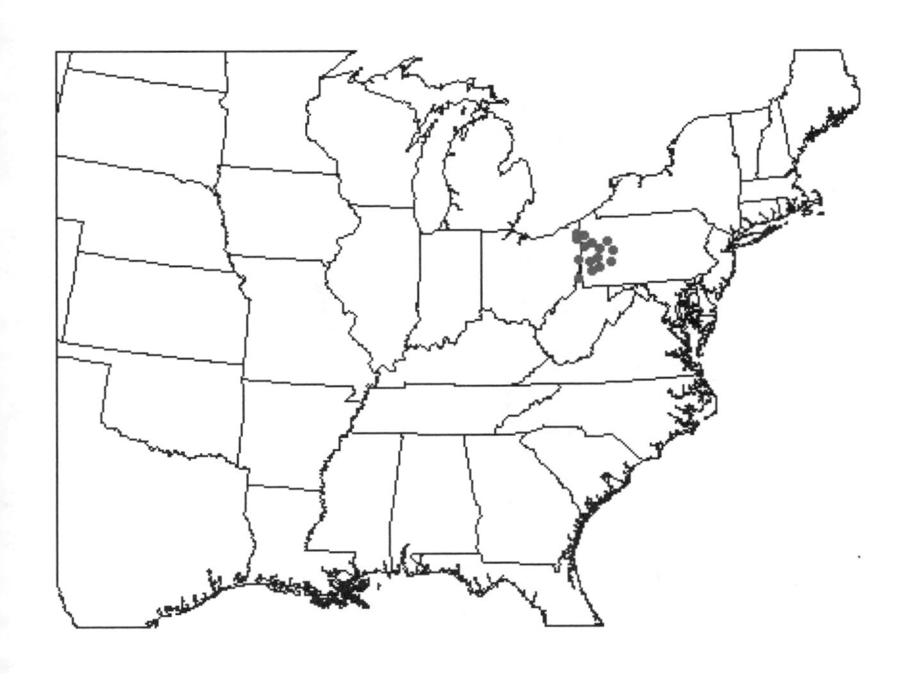

17-year cicada:  Brood VIII distribution.

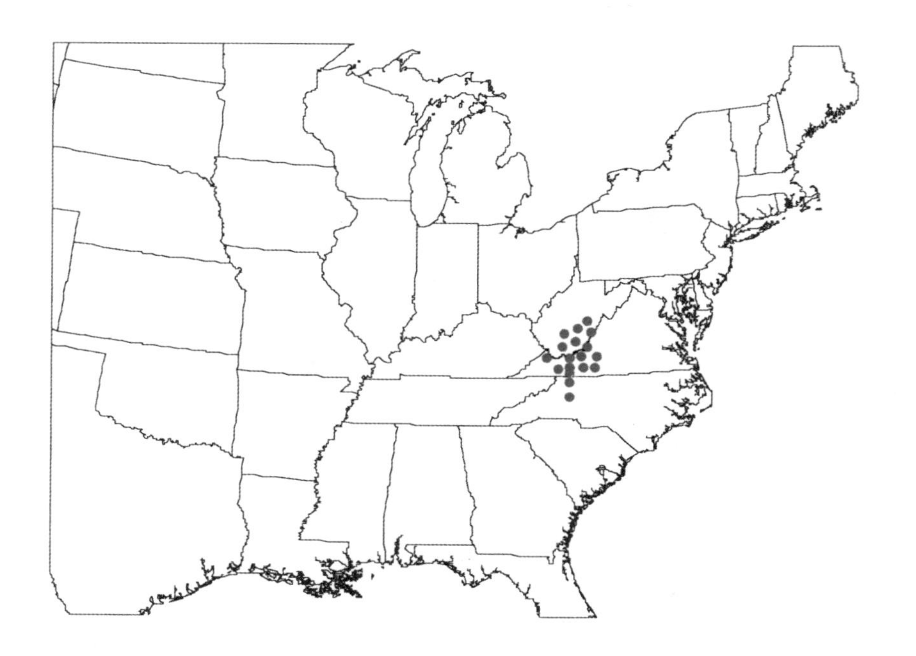

17-year cicada:  Brood IX distribution.

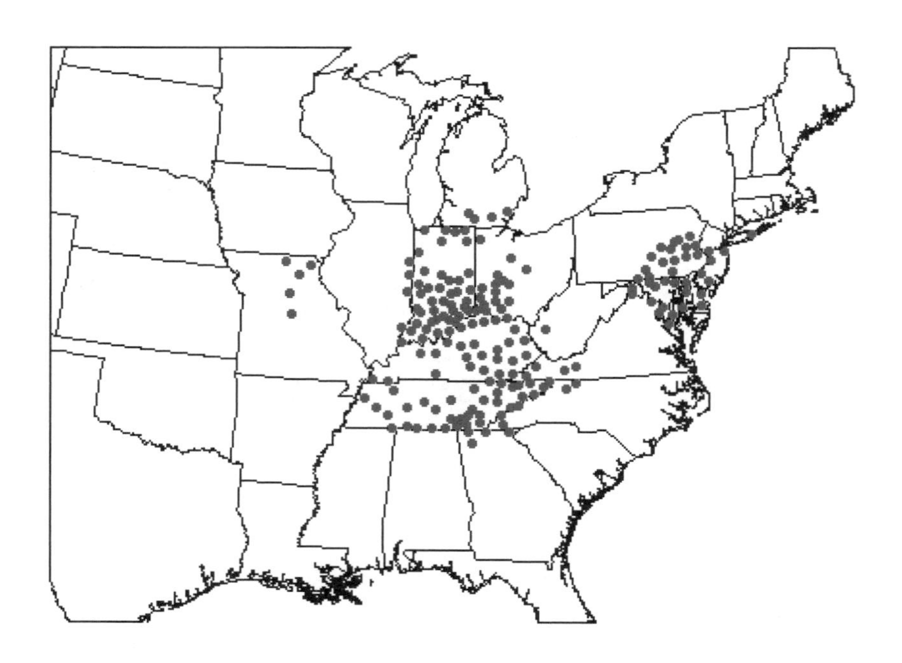

17-year cicada:  Brood X distribution.

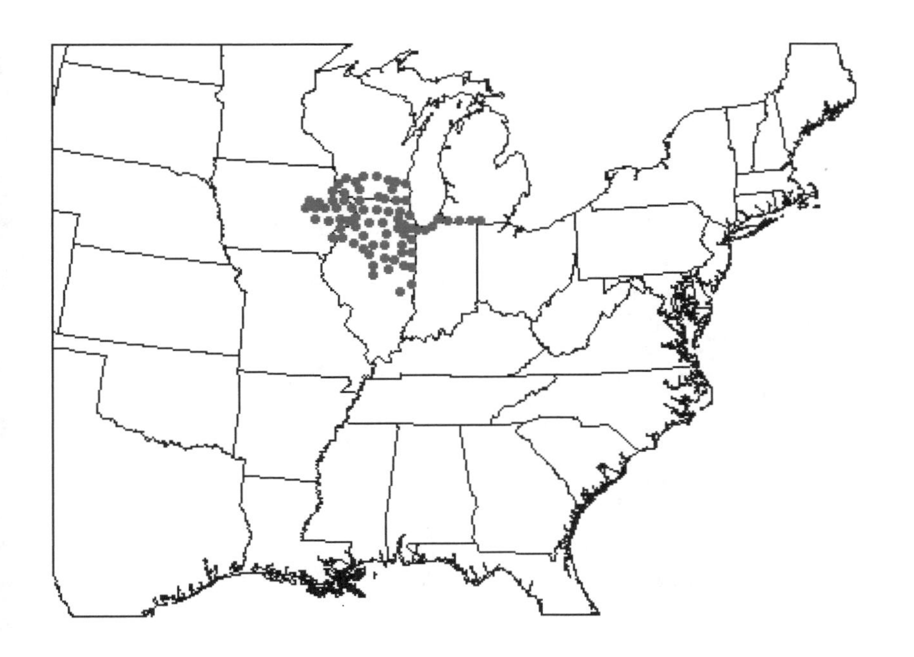

17-year cicada:  Brood XIII distribution.

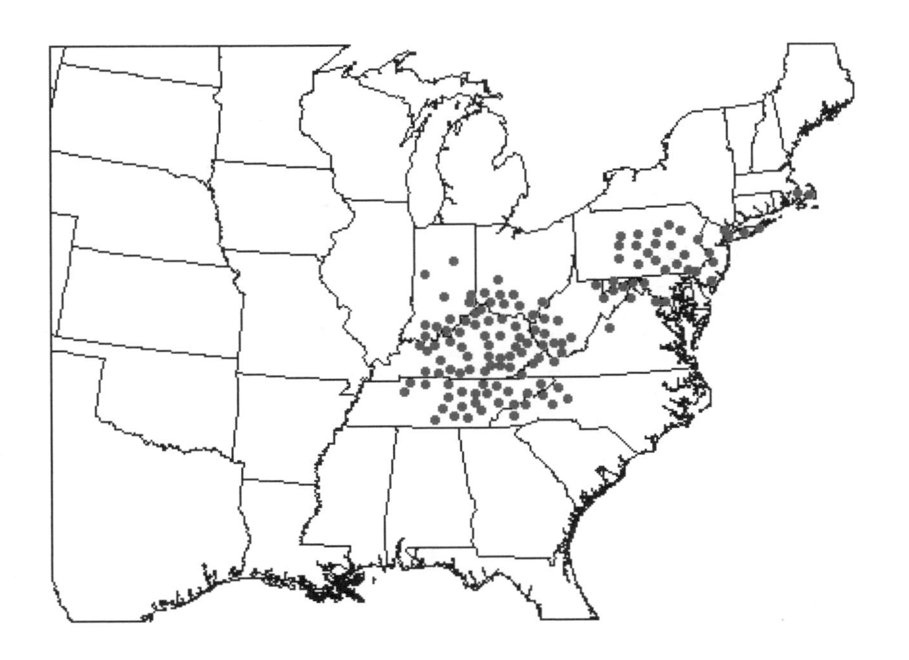

17-year cicada:  Brood XIV distribution.

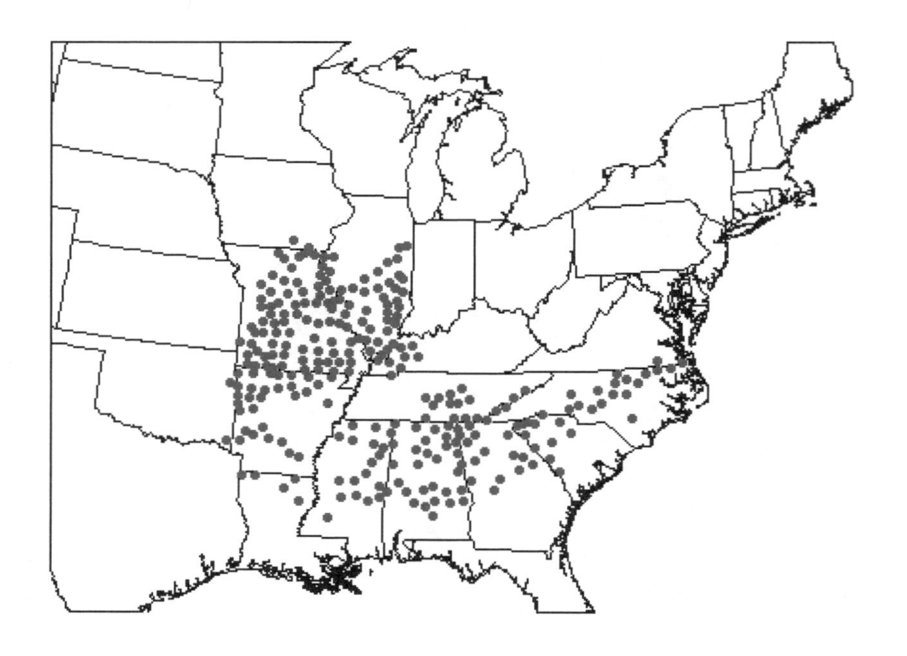

13-year cicada:  Brood XIX distribution.

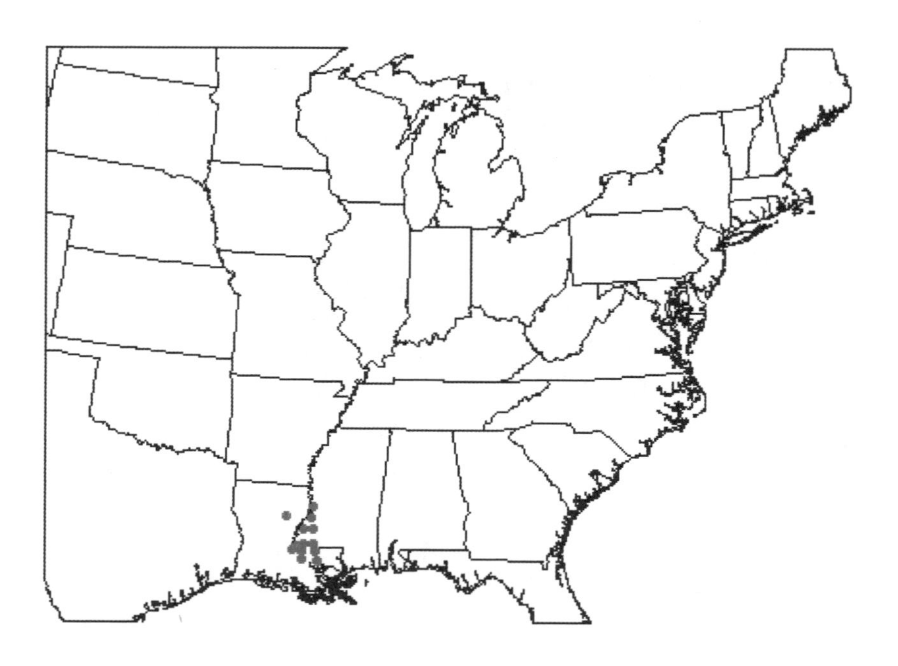

13-year cicada:  Brood XXII distribution.

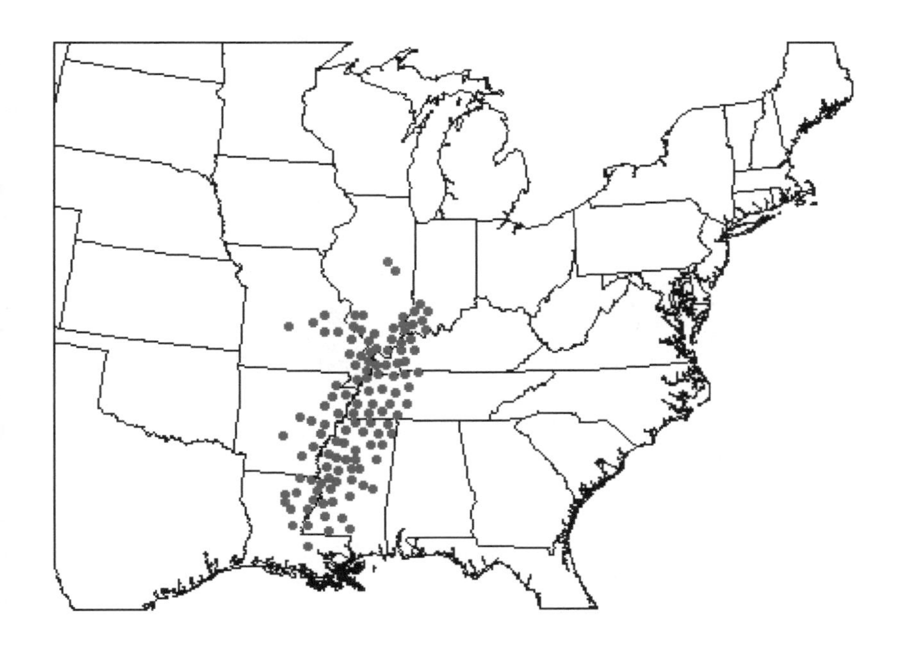

13-year cicada: Brood XXIII distribution.

# APPENDIX D. Professional Groups and Web Sites

**Philacicada Society**

A group founded by scientists who research the periodical cicadas. Its goal is to improve communication between periodical cicada specialists and others interested in periodical cicada research. For more information, contact Gene Kritsky, Treasurer, Philacicada Society, College of Mount St. Joseph, Cincinnati, OH 45233.

**Worldwide Web Sites:**

New sites on periodical cicadas and cicadas are being added to the World Wide Web all the time. To monitor these new sites, use your preferred search engine with the term "periodical cicada." Two sites that are of interest to Ohio's periodical cicada watchers are:

The College of Mount St. Joseph -
http://www.msj.edu/cicada/

The University of Michigan Museum of Zoology -
http://www.ummz.lsa.umich.edu/magicicada/

# APPENDIX E.
# Reading List and Selected References

## General Interest Books

Benson, Adolph B. 1964. Peter Kalm's travels in North America. Wilson-Erickson, Inc. New York, New York. 797 p.
Bradford, William. 1959. Of Plymouth Plantation 1620-1647. Alfred A. Knopf. New York, New York. 437 p.
Jefferson, Thomas. 1944. Thomas Jefferson's garden book. American Philosophical Society. Philadelphia, Pennsylvania. 704 p.
McCook, Henry C. 1904. The periodical cicada. Harpers' Magazine 109: 44-49.
McNeal, J. 1888. Insect Life 1: 50.
Smith, John B. 1894b. Insect Life 7: 192-195.

## Technical Books and Selected References for Chapters

Alexander, A. G. 1888. After-effect of the oviposition of the periodical cicada. Insect Life 1: 15.
Alexander, Richard D. and Thomas E. Moore. 1958. Studies on the acoustical behavior of seventeen cicadas (Hompotera: Cicadidae: *Magicicada*). The Ohio Journal of Science 58: 107-27.
Alexander, Richard D. and Thomas E. Moore. 1962. The evolutionary relationships of 17-year and 13-year cicadas, and three new species (Homoptera, Cicadidae, *Magicicada*). Miscellaneous Publications of the Museum of Zoology, University of Michigan. 121: 1-59.
Allard, H. A. 1937. Some observations on the behavior of the periodical cicada *Magicicada septendecim* L. American Naturalist 71: 588- 604.
Allard, H. A. 1946. Synchronous singing of the17-year cicada. Proceedings of the Entomological Society of Washington 48 (4): 93-95.
Alwood, W. B. 1902. What to do with locusts. Rural New Yorker May 17: 351.
Alwood, W. B. 1903. A note on the oviposition of the seventeen-year locust (*Cicada septendecim* Linn.) Bulletin No. 40, Division of Entomology, U. S. Department of Agriculture, August : 75-76.
Andrews, E. A. 1921. Periodical cicadas in Baltimore, Maryland. Scientific Monthly 12: 310-320.
Anonymous. 1881. The periodical cicada. Scientific American 45: 21.
Apgar, E. A. 1887. Some observations on the anatomy of *Cicada septendecim*. Journal of the Trenton Natural History Society January: 43-46.
Archie, James, Chris Simon, and D. Wartenberg. 1985. Geographic patterns and population structure in periodical cicadas based on spatial analysis of allozyme frequencies. Evolution 39 (6): 1261-1274.

Asquith, Dean. 1954. The periodical cicada in southern Pennsylvania in 1953. Journal of Economic Entomology 47 (2): 457-459.

Asquith, Dean. 1957. N-methyl-1-naphthyl carbamate as a killing agent for the periodical cicada. Journal of Economic Entomology 50 (5): 696.

Barnes, Harley. 1880. Seventeen-year cicada in Ohio. American Entomologist 3: 227-228.

Barnes, Harley. 1880. Periodical cicada in Geauga County, Ohio. American Entomologist 3: 226.

Bartram, John. 1804. Additional observations on *Cicada septendecim*. Barton Medical and Physical Journal 1: 56-59.

Bartram, Moses. 1796. Observations on the cicada, or locusts of America, which appears periodically once in 16 or 17 years, 1766. Annual Register or a View of the History, Politics, and Literature for the year 1767: 103.

Beamer, R. H. 1931. Notes on the 17-year cicada in Kansas. Journal of the Kansas Entomological Society 4: 53-58.

Bessey, C. E. 1880. On the distribution of the 17-year cicada of the brood of 1878, or Riley's brood XIII, in Iowa. American Entomologist 3: 27-30.

Bessey, C. E. 1883. The periodical cicada in southeastern Massachusetts. American Naturalist 17: 1071.

Bethune, C. J. S. 1875. Grasshoppers or locusts. Annual Report of the Entomological Society of Ontario for 1874 : 29.

Boyd, W. M. 1952. A premature emergence of periodical cicadas. Journal of the New York Entomological Society 60: 156.

Britton, W. E. 1903. Third report of the state entomologist. Report of the Connecticut Agricultural Experiment Station for 1903: 214.

Brown, J. J. and G. M. Chippendale. 1973. Nature and fate of the nutrient reserves of the periodical (17-year) cicada. Journal of Insect Physiology 19: 607-614.

Bryce, David and Nevin Aspinwall. 1975. Sympatry of two broods of the periodical cicada (*Magicicada*) in Missouri. The American Midland Naturalist 93 (2): 405-454.

Buckhout, William A. 1890. The periodical cicada in Pennsylvania. Report of the Pennsylvania Agricultural Experiment Station for 1889: 182-187.

Bulmer, M. G. 1977. Periodical insects. American Naturalist 111: 1099-1117.

Burnett, Dr. W. J. 1851a. Proceedings of the Boston Society of Natural History 4: 71, 111.

Burnett, Dr. W. J. 1851b. Points in the economy of the 17-year locust (*Cicada septendecim*) bearing on the plural origin and special local creation of the species. Proceedings of the American Association for the Advancement of Science, 6th Meeting: 307-311.

Butler, Amos W. 1886. The periodical cicada in southeastern Indiana. Bulletin No. 12, Division of Entomology, U. S. Department Agriculture (July 13): 24-31.

Cassin, John. 1851. Notes on the above species of cicada (*C. cassini*), and on the *Cicada septendecim* Linn. Proceedings of the Academy of Natural Science, Philadelphia 5: 273-275.

Chambers, V. T. 1880. American Entomologist 3: 77.

Children, J. G. 1837. Proceedings of the Entomological Society of London 1: xxx.

Collinson, Peter. 1764. Some observations on the cicada of North America. Philosophical Transactions of the Royal Society of London 54: 65-69.

Cook, Albert J. 1868. Remarks on some insects injurious to vegetation in Michigan. 7th Annual Report of the Secretary of the State Board Agriculture Michigan for 1868: 163-170.
Cooper, Kenneth W. 1941. *Davispia bearcreekensis* Cooper, a new cicada from the Paleocene, with a brief review of the fossil Cicadidae. American Journal of Science 239 (4): 286-304.
Cory, Ernest H. and Paul Knight. 1937. Observations on brood X of the periodical cicada in Maryland. Journal of Economic Entomology 30 (2): 287-294.
Cox, Randall T. and C. E. Carlton. 1988. Paleoclimatic influences in the ecology of periodical cicadas (Insecta: Homoptera: Cicadidae: *Magicicada* spp.). The American Midland Naturalist 120 (1): 183-193.
Cox, Randall T. and C. E. Carlton. 1991. Evidence of genetic dominance of the 13-year life cycle in periodical cicadas (Homoptera: Cicadidae: *Magicicada* spp.). The American Midland Naturalist 125: 63-74.
Craig, F. Waldo. 1941. Observations on the periodical cicada. Journal of Economic Entomology 34: 122-23.
Craig, F. Waldo. 1936. Economic control of *Magicicada* (*Tibician*) *septendecim* L. in a grove of century-old oaks. Journal of Economic Entomology 29 (1): 190-192.
Craig, F. Waldo. 1941. Observations on the periodical cicada. West Virginia State Department of Entomology 34 (1): 122-123.
Cuthbert, Nicholas. L. and Mabel J. Cuthbert. 1945. A cat that eats cicadas. Entomological News 56 (6): 143.
Davis, J. F. 1819. On the *Cicada septendecim*. Journal of Science and Arts of the Royal Institute 6: 372-374.
Davis, J. J. 1953. Pehr Kalm's description of the periodical cicada, *Magicicada septendecim* L. The Ohio Journal of Science 53 (3): 138-142.
Davis, J. J. 1954. Insects of Indiana in 1953. Proceedings of the Indiana Academy of Science 63: 152-156.
Davis, Millard C. 1963. The voice of the cicada. American Biology Teacher 25 (1): 21-23.
Davis, William T. 1885. The periodical cicada on Staten Island. American Entomologist 1 (5): 91.
Davis, William T. 1894a. The harvest flies (cicada) of Staten Island, N. Y. American Naturalist 28 (328): 363-364.
Davis, William T. 1894b. The 17-year locust on Staten Island. American Naturalist 28 (329): 452.
Davis, William T. 1894c. The 17-year locust on Staten Island in 1894. Proceedings of the Natural Science Association of Staten Island 4 (9): 33-35.
Davis, William T. 1894d. The 17-year locust on Staten Island. Proceedings of the Natural Science Association of Staten Island 4 (4): 13-15.
Davis, William T. 1894e. The 17-year cicada on Staten Island. Journal of the New York Entomological Society 2 (1): 38-39.
Davis, William T. 1925. *Cicada tibicen*, a new South American species, with records and descriptions of North American cicadas. Journal of the New York Entomological Society 33: 35-51.
Davis, William T. 1926. The cicadas or harvest flies of New Jersey. New Jersey Department Agriculture Bureau of Statistics and Inspection Circular 97: 3-27.

Davis, William T. 1928. The occasional appearance of the seventeen year cicada in the fall, and brood no. I on Long Island, N. E., in 1927. Bulletin of the Brooklyn Entomological Society 23 (2): 64-66.

Dean, R. W. 1963. Effect of soil applications of nematocides on emergence of periodical cicada. Journal of Economic Entomology 56 (4): 540.

Deay, Howard O. 1953. The periodical cicada *Magicicada septendecim* (L) in Indiana. Proceedings of the Indiana Academy of Science 62: 203-206.

Deay, Howard O. 1953. The periodical cicada *Magicicada septendecim* (L) in Indiana. Indiana Academy of Science: 203-206.

Dimmock, G. 1872a. Insects infesting apple trees. New England Homestead 5 (4): 25.

Dimmock, G. 1872b. Insects infesting apple trees. New England Homestead 5 (7): 49.

Dun, Walter A. 1886. Notes on the occurrence of the periodical cicada. Journal of the Cincinnati Society of Natural History 8 (4): 233-234.

Dunning, D. Covalt, John A. Byers, and C. Dewitt Zanger. 1979. Courtship in two species of periodical cicadas, *Magicicada septendecim* and *M. cassini*. Animal Behavior 27: 1073-1090.

Dwight, Timothy. 1823. Travels in New-England and New-York. William Baynes and Son, Edinburgh 1: 30-31.

Dybas, Henry S. and Monte Lloyd. 1962. Isolation by habitat in two synchronized species of periodical cicadas (Homoptera: Cicadidae: *Magicicada*). Ecology 43: 444-459.

Dybas, Henry S. and Monte Lloyd. 1974. The habitats of 17-year periodical cicadas (Homoptera: Cicadidae: *Magicicada* spp.). Ecological Monographs 44 (3): 279-324.

Dybas, H. S. and D. D. Davis. 1962. A population census of seventeen-year periodical cicadas (Homoptera: Cicadidae: *Magicicada*). Ecology 43: 432-443.

Evans, Gurdon. 1852. Insects injurious to vegetation. Transactions of the New York State Agricultural Society for 1851 11: 741-751.

Fabricius, Johann C. 1775. Systema entomological. Flensbvrgi et Lipsiae. 832 p.

Felt, Ephraim P. 1899. Notes of the year for New York. Country Gentleman September 14: 733.

Felt, Ephraim P. 1899. Notes of the year for New York. Bulletin Number X, new series, Entomological Division , U. S. Department of Agriculture, November 20: 62.

Felt, Ephraim P. 1900. Fifteenth report of the state entomologist on the injurious and other insects of the state of New York. Bulletin of the New York State Museum 6 (31): 544.

Felt, Ephraim P. 1901. Seventeen-year cicada. Country Gentleman November 7: 902.

Felt, Ephraim P. 1902. Spraying for cicada. Country Gentleman March 13: 219.

Felt, Ephraim P. 1903. Eighteenth report of the state entomologist on injurious and other insects of the state of New York. Bulletin of the New York State Museum, Albany 64: 113.

Fernald, C. H. and H. T. Fernald. 1901. Report of the entomologists. Thirteenth Annual Report, Hatch Experiment Station, Massachusetts, January: 86.

Fisher, Dr. J. C. 1851. On a new species of cicada. Proceedings of the Academy of Natural Science Philadelphia 5: 272-273.

Fitch, Asa. 1854. Report on the noxious, beneficial, and other insects of the State of New York. Transactions of the New York State Agricultural Society for 1854 (14): 742-753.

Fitch, Asa. 1860. The entomologist, Number 22 - The 17-year cicada. The Country Gentleman 15: 210.
Forbes, Stephen A. 1898. The seventeen-year cicada. Prairie Farmer June: 9.
Forbush, E.H. 1924. Gulls and terns feeding on the 17-year cicada. Auk 41: 468-470.
Forsley, C. G. 1846. On the *Cicada septendecim* in 1835 in Louisiana. Proceedings of the Boston Society of Natural History 2: 162.
Forsythe, H. Y., Jr. 1966. Screening insecticides for control of the adult periodical cicada. Journal of Economic Entomology 59: 1413-1416.
Forsythe, H. Y., Jr. 1969. The 17-year cicada in 1968. (*Tibicen septendecim*). Pesticide News 22: 54.
Forsythe, H. Y., Jr. 1975. Laboratory and field evaluation of insecticides for control of periodical cicadas (*Magicicada septendecim*) in orchards. Research Circular of the Ohio Agricultural Research and Development Center 211: 1-9.
Forsythe, H. Y., Jr. 1976a. Estimating nymphal populations of the 17-year cicadas in eastern Ohio, 1968. The Ohio Journal of Science 76: 95-96.
Forsythe, H. Y., Jr. 1976b. Number of seventeen-year cicada eggs per nest. Environmental Entomology 5: 169-170.
Forsythe, H. Y., Jr. 1976c. Distribution and species of 17-year cicadas in broods V and VIII in Ohio. The Ohio Journal of Science 76 (6): 254-258.
Forsythe, H. Y., Jr. 1977. Effect of sun-exposure on emergence of 17 year cicadas. The Ohio Journal of Science 77: 183-185.
Fries, Adelaide L., Editor. 1968. Records of the Moravians in North Carolina. Publication of the North Carolina Historical Commission 2: 583.
Fries, Adelaide L., Editor. 1970. Records of the Moravians in North Carolina. Publication of the North Carolina Historical Commission 5: 2361.
Garman, H. 1903. Seventeen-year locusts in Kentucky. Bulletin of the Kentucky Agricultural Experiment Station 107: 81-100.
Garman, H. 1905. The seventeen-year locust will not appear in Kentucky this year (1905). Kentucky Agricultural Experiment Station 120: 74-76.
Gassard, Harry A. 1917. Distribution of the Ohio broods of periodical cicada with reference to soil. Bulletin of the Ohio Agricultural Experiment Station (311): 555-577.
Glover, Townend. 1867. Report (U.S.) Commission of Agriculture for 1866: 29.
Glover, Townend. 1868. Report (U. S.) Commission of Agriculture for 1867: 67-71.
Glover, Townend. 1873. Report of the entomologist and curator of the museum. Report (U.S.) Commission of Agriculture for 1872: 112-138.
Goldstein, Bessie. 1929. A cytological study of the fungus *Massospora cicadina*, parasitic on the 17-year cicada, *Magicicada septendecim*. American Journal of Botany 16 (6): 394-401.
Graham, C. and A. B. Cochran. 1954. The periodical cicada in Maryland in 1953. Journal of Economic Entomology 47: 242-244.
Graham, Castillo and Elroy R. Krestensen. 1957. A residual spray for control of the periodical cicada. Journal of Economic Entomological 50 (6): 713-715.
Hahn, Walter L. 1905. Note on the recurrence of brood V of *Tibicen septendecim* in Porter County, Indiana, during 1905. Proceedings of the Indiana Academy of Science: 227.
Hahus, S. C. and K. G. Smith. 1990. Food habits of *Blarina*, *Peromyscus*, and *Microtus* in relation to an emergence of periodical cicadas (*Magicicada*). Journal of Mammalogy 71: 249-252.

Hamilton, D. E. and M. L. Cleveland. 1963. Periodical cicadas in 1963, brood 23. Proceedings of the Indiana Academy of Science 73: 167-170.

Harris. T. W. 1852. Insects of New England 1852: 180-189.

Hartzell, Albert. 1954. Periodical cicada. Contribution of the Boyce Thompson Institute 17 (6): 375-377.

Haseman, Leonard. 1915. The periodical cicada in Missouri. University of Missouri Agricultural Experiment Station Bulletin 137. 33 p.

Heath, James Edward. 1967. Temperature responses of the periodical "17-year" cicada, *Magicicada cassini* (Homoptera, Cicadidae). The American Midland Naturalist 77 (1): 64-76.

Herrick. E. C. 1862. Uprising of the 17-year cicada in New Haven County, Conn., in 1860. American Journal of Arts and Science, 2d series 33: 433-434.

Hewitt, G. M., R. A. Nichols, and M. G. Ritchie. 1988. 1868 and all that for *Magicicada*. Nature 336: 206-207.

Hickernell, L.M. 1920. The digestive system of the periodical cicada, *Tibicen septendecim* Linn. I. Morphology of the system in the nymph. Annals of the Entomological Society of America 13: 223-242.

Hickernell, L.M. 1923. The digestive system of the periodical cicada, *Tibicen septendecim* Linn. III. Morphology of the system in the nymph. Biological Bulletin 45: 213-221.

Hildreth, S. P. 1826. American Journal of Science and Arts 10: 327-329.

Hildreth, S. P. 1830. Notices and observations on the American cicada or locust. American Journal of Science and Arts 18: 47-50.

Hildreth, S. P. 1847a. *Cicada septendecim* in 1846. American Journal of Science and Arts 3: 216-218.

Hildreth, S. P. 1847b. *Cicada septendecim* in 1846. Annals and Magazine of Natural History 20: 136-138.

Hogmire, W. H., T. A. Baugher, V. L. Crim, and S. I. Walter. 1990. Effects and control of periodical cicada (Homoptera: Cicadidae) oviposition injury on nonbearing apple trees. Journal of Economic Entomology 83: 2401-2404.

Hoopes, Abner. 1902. Entomological News 18: 108-109.

Hopkins, A. D. 1898. The periodical cicada in West Virginia. Bulletin of the West Virginia Agricultural Experiment Station 50: 1-46.

Hopkins, A. D. 1898. Some notes on observations in West Virginia. Bulletin, new series, Division of Entomology, U. S. Department of Agriculture 17: 44-49.

Hopkins, A. D. 1900. The periodical cicada or seventeen-year locust in West Virginia. Bulletin Number 68, West Virginia Agricultural Experiment Station : 259-330.

Hopkins, A. D. 1901. Circular of warning. West Virginia Agricultural Experiment Station. October 15.

Hopkins, A. D. 1902. What to do with locusts. Rural New Yorker May 17: 351.

Hoppensteadt, F. C. and J. B. Keller. 1976. Synchronization of periodical cicada emergences. Science 194: 335-337.

Howard, Leland O. 1886. Proceedings of the Entomological Society of Washington 1: 29.

Howard, Leland O. 1898. A new egg-parasite of the periodical cicada. Canadian Entomologist 30: 102-103.

Howard, W. J. 1937. Bird behavior as a result of emergence of seventeen year locusts. Wilson Bulletin 49: 43-44.

Hunter, P.E. and H.O. Lund. 1960. Biology of the periodical cicada in Georgia. Journal of Economic Entomology 53: 961-963.
Hunter, W. D. 1902. The periodical cicada in 1902. Division of Entomology, U. S. Department of Agriculture, Circular Number 44, new series, March: 4.
Hyatt, J. D. 1896. *Cicada septendecim*; its mouthparts, and armor. American Monthly Microscopical Journal: 46.
Hyslop, J. A. 1935. The periodical cicada. U.S. Department of Agriculture Bureau of Entomological and Plant Quarantine 17 p.
Hyslop, J. A. 1940. The periodical cicada, brood XIV. U.S. Department of Agriculture Bureau of Entomological and Plant Quarantine 13 p.
Jacobs, Merle. 1954. Observations on the two forms of the periodical cicada *Magicicada septendecim*. Proceedings of the Indiana Academy of Science 63: 177-179.
James, D. A., Kathy S. Williams, and K. G. Smith. 1986. Survey of 1985 periodical cicada (Homoptera: *Magicicada*) emergence sites in Washington County, Arkansas, with reference to ecological implications. Proceedings of the Arkansas Academy of Science 40: 37-39.
Johnson, W. G. 1901. Timely warning to fruit growers. American Agriculturist July 13: 32.
Josselyn, John. 1865. An account of two voyages to New-England. William Veazie. Boston. 211 p.
Kalm, Pehr. 1771. Travels into North America. W. Eyres. Warrington, England. 3 vols.
Kalm, Pehr. 1778. Three year's travels through the interior parts of North-America, for more than five thousand miles. Key & Simpson. Philadelphia: 325-327.
Kalm, Pehr. 1756. Beskrifning pa et slagts Gras-Hoppor, uti Norra Americas (*Cicada septendecim*). Vetensk. Academy Handl. 17: 101-116.
Karban, Richard. 1980. Periodical cicada nymphs impose periodical oak tree wood accumulation. Nature 287: 326-327.
Karban, Richard. 1981a. Effects of local density on fecundity and mating speed for periodical cicadas. Oecologia (Berlin) 51: 260-264.
Karban, Richard. 1981b. Flight and dispersal of periodical cicadas. Oecologia 49: 385-390.
Karban, Richard. 1982a. Experimental removal of 17-year cicada nymphs and growth of host apple trees. Journal of the New York Entomological Society 90: 74-81.
Karban, Richard. 1982b. Increased reproductive success at high densities and predator satiation for periodical cicadas. Ecology 63: 321-328.
Karban, Richard. 1983. Induced responses of cherry trees to periodical cicada oviposition. Oecologia 59: 226-231.
Karban, Richard. 1984. Opposite density effects of nymphal and adult mortality for periodical cicadas. Ecology 65: 1656-1661.
Karban, Richard. 1985. Addition of periodical cicada nymphs to an oak forest: effects on cicada density, acorn production, and rootlet density. Journal of the Kansas Entomological Society 58: 269-276.
Karban, Richard. 1986. Prolonged development in cicadas. Pages 222-235 *in* The evolution of insect life cycles. F. Taylor and Richard Karban, editors. Springer-Verlag. New York, New York. 287 p.
Karlin, A. A., Kathy S. Williams, K. G. Smith, and D. W. Sugden. 1991. Biochemical evidence for rapid changes in heterozygosity in a population of periodical cicadas (*Magicicada tredecassini*). The American Midland Naturalist 125: 213-221.

Kite, William. 1870. The 17-year cicada. American Naturalist 3: 106.

Knepp, T. H. 1963. The 1962 appearance of brood II of the seventeen year cicada in Monroe Co., Pa. Pan-Pacific Entomologist 40 (4): 222-226.

Kohl, W. M. 1891. Seventeen-year locust. Farmer's Review November 4: 700.

Kritsky, Gene. 1988a. The 1987 emergence of the periodical cicada (Homoptera: Cicadidae: *Magicicada* spp.: Brood X) in Ohio. The Ohio Journal of Science 88: 168-170.

Kritsky, Gene. 1988b. An historical analysis of periodical cicadas in Indiana (Homoptera: Cicadidae). Proceedings of the Indiana Academy of Science 97: 295-322.

Kritsky, Gene. 1989. Periodical cicadas. Nature 341 (6240): 288-289.

Kritsky, Gene. 1992. The 1991 emergence of the periodical cicadas in Ohio. The Ohio Journal of Science 92 (1): 38-39.

Kritsky, Gene and Ronald H. Meyer. 1976. The emergence of the periodical cicada (Brood XXIII) in Illinois in 1976 (Homoptera: Cicadidae). Transactions of the Illinois State Academy Science 69 (2): 196-199.

Kritsky, Gene and Sue Simon. 1996. The unexpected 1995 emergence of periodical cicadas (Homoptera: Cicadidae: *Magicicada* spp.) in Ohio. The Ohio Journal of Science 96: 27-28.

Kritsky, Gene and Frank N. Young. 1990. Observations of periodical cicadas (Brood XXIII) in Indiana in 1989 (Homptera: Cicadidae). Proceedings of the Indiana Academy of Science 99: 25-28.

Kritsky, Gene and Frank N. Young. 1991. Observations on periodical cicadas (Brood XIII) in Indiana in 1990. Proceedings of the Indiana Academy of Science 100: 45-47.

Kritsky, Gene and Frank N. Young. 1992. Observations on periodical cicadas (Brood XIV) in Indiana in 1991 (Homptera: Cicadidae). Proceedings of the Indiana Academy of Science 101: 59-61.

Krom, Stephen H. 1894. The hut-building cicada. Scientific American November 10: 295.

Lander, Benjamin. 1894a. Cicada hut-builders. Scientific American November 24: 327.

Lander, Benjamin. 1894b. Hut-building 17-year cicadas. Scientific American October 24: 233-234.

Lander, Benjamin. 1895. Domed burrows of cicada *septendecim*. Journal of the New York Entomological Society III: 33-38.

Lander, Benjamin. 1899. Note on the seventeen-year cicada. Journal of the New York Entomological Society, vii: 212-214.

LeBaron, William. 1871a. Locust or periodical cicada. Prairie Farmer April: 42.

LeBaron, William. 1871b. Prairie Farmer June: 42.

LeBaron, William. 1872. Second Report on the Insects in Illinois: 121-133.

Leidy, Dr. Joseph. 1867. Proceedings of the Academy of Natural Science, Philadelphia 21: 93.

Leidy, Dr. Joseph. 1851. Proceedings of the Academy of Natural Science, Philadelphia 5: 235.

Leidy, Dr. Joseph. 1877. Remarks on the 17-year locust, etc. Proceedings of the Academy of Natural Science, Philadelphia 31: 260-261.

Leonard, David E. 1964. Biology and ecology of *Magicicada septendecim* (L.) (Hemiptera: Cicadidae). Journal of the New York Entomological Society 72 (1): 19-23.
Lin, Norman. 1965. The use of emergence holes of the cicada killer (*Sphectus speciosus*) as nest burrows by *Tachytes* (Hymenoptera: Sphecidae). Bulletin of the Brooklyn Entomological Society 59/60: 82-84.
von Linné, Carl. 1758. Systema Naturae. 10th ed. 823 p.
Lintner, Joseph A. 1882. The 17-year locust. Ontario [N.Y.] County Times 28: 3.
Lintner, Joseph A. 1885a. The 13-year cicada. Argus (Albany) October 11: 4.
Lintner, Joseph A. 1885b. The 17-year locust, etc. Second report on the injurious and other insects of the State of New York. Albany (February, 1886): 167-179.
Lintner, Joseph A. 1887a. The 17-year locust, *Cicada septendecim*. The Owl 2: 17-19.
Lintner, Joseph A. 1887b. An experiment with the 13-year cicada. 39th Annual Report of the State Museum of Natural History for 1885 39: 111-112.
Lintner, Joseph A. 1889. An experiment with the 13-year cicada. Fifth Report on the Injurious and other Insects of the State of New York: 276-278.
Lintner, Joseph A. 1891. *Cicada septendecim* Linn. The periodical cicada. Seventh Report on the Injurious and other Insects of the State of New York: 296-301.
Lintner, Joseph A. 1893. The periodical cicada. Country Gentleman March 23: 226.
Lintner, Joseph A. 1894. The periodical cicada, or the 17-year locust. Circular, Albany, New York. June 19: 4.
Lintner, Joseph A. 1898. The periodical cicada. 12th Annual Report of the State Entomologist of New York for 1896 (May, 1898) 12: 272-289.
Lloyd, Monte. 1966. The periodical cicada problem. II. Evolution. Evolution 20 (4): 466-505.
Lloyd, Monte. 1984. Periodical cicadas. Antenna 8 (2): 79-91.
Lloyd, Monte. 1987. A successful rearing of 13- year periodical cicadas beyond their present range and beyond that of 17-year cicadas. The American Midlland Naturalist 117: 362-368.
Lloyd, Monte and Henry S. Dybas. 1966. The periodical cicada problem. I. Population ecology. Evolution 20 (2): 133-149.
Lloyd, Monte and JoAnn White. 1976. Sympatry of periodical cicada broods and the hypothetical four-year acceleration. Evolution 30: 786-801.
Lloyd, Monte and JoAnn White. 1976. On the oviposition habits of 13-year versus 17-year periodical cicadas of the same species. Journal of the New York Entomological Society 84 (3): 148-155.
Lloyd, Monte and JoAnn White. 1987. Xylem feeding by periodical cicada nymphs on pine and grass roots, with novel suggestions for pest control in conifer plantations and orchards. The Ohio Journal of Science 87 (3): 50-54.
Lloyd, Monte and JoAnn White. 1983. Why is one of the periodical cicadas (*Magicicada septendecula*) a comparatively rare species? Ecological Entomology 8: 293-303.
Lloyd, Monte and Richard Karban. 1983. Chorusing centers of periodical cicadas. Journal of the Kansas Entomological Society 56 (3): 299-304.
Lloyd, Monte, Gene Kritsky, and Chris Simon. 1983. A simple Mendelian model for 13- and 17-year life cycles of periodical cicadas, with historical evidence of hybridization between them. Evolution 37 (6): 1162-1180.
Lloyd, Monte, JoAnn White, and N. Stanton. 1982. Dispersal of fungus-infected periodical cicadas to new habitat. Environmental Entomology 11: 852-858.

Long, Charles A. 1993. Evolution in periodical cicadas: a genetical explanation. Evolutionary Theory 10: 209-220.

Love, E. G. 1895. Notes on the 17-year cicada, *Cicada septendecim*. Journal of the New York Microscopical Society 11: 37-46.

Lowe, V. H. 1902. Miscellaneous notes on injurious insects. Bulletin No. 212, New York State Agricultural Experiment Station, April: 1-15.

Luken, J. O. and P. J. Kalisz. 1989. Soil disturbance by the emergence of periodical cicadas. Soil Science Society of America Journal 53: 310-313.

Maier, Chris T. 1980. A mole's-eye view of seventeen-year periodical cicada nymphs, *Magicicada septendecim* (Hemiptera: Homoptera: Cicadidae). Annals of the Entomological Society of America 73: 147-152.

Maier, Chris T. 1982a. Abundance and distribution of the seventeen-year periodical cicada, *Magicicada septendecim* Linnaeus (Hemiptera: Cicadidae-Brood II) in Connecticut. Proceedings of the Entomological Society of Washington 84 (3): 430-439.

Maier, Chris T. 1982b. Observations on the seventeen-year periodical cicada, *Magicicada septendecim* (Hemiptera: Homoptera: Cicadidae). Annals of the Entomological Society of America 75: 14-23.

Maier, Chris T. 1985. Brood VI of 17-year periodical cicadas, *Magicicada* spp. (Hemiptera: Cicadidae): New evidence from Connecticut, the hypothetical 4- year deceleration and the status of the brood. Journal of the New York Entomological Society 93 (2): 1019-1026.

Manter, J. A. 1974. Brood XI of the periodical cicada seems doomed. 25th Anniversary Mem. of the Connecticut Entomological Society 974: 100-101.

March, J. 1889. Wisconsin letter on *Cicada septendecim*. Insect Life 1: 218.

Marlatt, Charles L. 1895. The Hemipterous mouth. Proceedings of the Entomological Society Washington 3 (4): 241-249.

Marlatt, Charles L. 1898a. The periodical cicada. Bulletin of the U. S. Department of Agriculture, Division of Entomology. Number 14, new series. 148 p.

Marlatt, Charles L. 1898b. A new nomenclature for the broods of the periodical cicada. Some Miscellaneous Results of the Work of the Division of Entomology. Bulletin Number 18, new series: 52-58.

Marlatt, Charles L. 1898c. A consideration of the validity of the old records bearing on the distribution of the broods of the periodical cicada, with particular reference to the occurrence of broods VI and XXIII in 1898. Some Miscellaneous Results of the Work of the Division of Entomology. Bulletin Number 18, new series: 59-78.

Marlatt, Charles L. 1903a. Notes on the periodical cicada in the District of Columbia in 1902. Proceedings of the Entomological Society of Washington 5 (2): 124-126.

Marlatt, Charles L. 1903b. An early record of the periodical cicada. Proceedings of the Entomological Society of Washington 5 (2): 126-127.

Marlatt, Charles L. 1906. The periodical cicada in 1906. U. S. Department of Agriculture, Bureau of Entomology Circular 74: 1-5.

Marlatt, Charles L. 1907. The periodical cicada. Bulletin of the U. S. Department of Agriculture, Division of Entomology Number 71. 181p.

Marlatt, Charles L. 1908. A successful seventeen-year breeding record for the periodical cicada. Proceedings of the Entomological Society of Washington 9: 16-19.

Martin, Archie and Chris Simon. 1990a. Temporal variation in insect life cycles. BioScience 40 (5): 359-367.

Martin, Archie and Chris Simon. 1990b. Differing levels of among-population divergence in the mitochondrial DNA of periodical cicadas related to historical biogeography. Evolution 44: 1066-1080.
Martin, Archie and Chris Simon. 1988. Anomalous distribution of nuclear and mitochondrial DNA markers in periodical cicadas. Nature 336: 237-239.
Matthews, Thomas. 1705. Stedman's Library of American Literature 1: 462-463.
May, Robert M. 1979. Periodical cicadas. Nature 277: 347-349.
McCarthy, G. 1893. The periodical cicada. North Carolina Agricultural Experiment Station Bulletin 92: 108-109.
McCutchen, A. R. 1870. Periodical cicadas in Georgia. American Entomologist and Botanist 2: 372.
Meek, W. J. 1903. On the mouth parts of the Hemiptera. Bulletin of the University of Kansas 2: 257-277.
Mill, Alfred S. 1929. Periodical cicada observations. Journal of Economic Entomology 22 (3): 594.
Moreton, Nathaniel. 1669. "New England's Memorial." Cambridge, England. 198 p.
Morris, John G. 1870. Seventeen-year locust two years too late. American Entomologist and Botanist 2: 304.
Morris, M. H. 1846. Proceedings of the Academy of Natural Science, Philadelphia 3: 132-134.
Morris, M. H. 1847a. Apple and pear trees destroyed by the locust. American Agriculturist 6: 86-87.
Morris, M. H. 1847b. Proceedings of the Academy of Natural Science, Philadelphia 3: 190-191.
Morris, M. H. 1848. Destruction of fruit trees by the seventeen-year locust. American Agriculture 7: 279.
Morton, S. G., *et al.* 1843. Proceedings of the Academy of Natural Science, Philadelphia 1: 277-280.
Murtfeldt, M. E. 1889. Report U. S. Department of Agriculture 1888: 135.
Nixon, P. L. 1990. Periodical cicadas to emerge in northern Illinois. Illinois Natural History Survey Reports (297): 3-4.
Oldenburg, Henry. 1666. Some observations of swarms of strange insects and the mischiefs done by them. Philosophical Transactions of the Royal Society of London 1 (8): 137
Osborn, Herbert. 1878. The 17-year locust. Western Farm Journal. July.
Osborn, Herbert. 1879a. Report of noxious insects. Transactions of the Iowa State Horticultural Society for 1878 13: 368-402.
Osborn, Herbert. 1879b. Seventeen-year locusts. College Quarterly 2: 58.
Osborn, Herbert. 1882. Insects of the forest—*Cicada septendecim.* Iowa State Leader December 2.
Osborn, Herbert. 1884. Insects of the orchard. Bulletin of the Iowa Agricultural College (2): 87-97.
Osborn, Herbert. 1889. Notes on destructive insects. Annual Report of the Iowa State Agricultural Society for 1888: 670-680.
Osborn, Herbert. 1896. Observations on the Cicadae of Iowa. Proceedings of the Iowa Academy of Science, III, 1895: 194-203.
Osborn, Herbert. 1902a. Insects affecting forest trees. Proceedings of the Columbus Horticultural Society 17: 79-92.

Osborn, Herbert. 1902b. Some notable insect occurrences in Ohio for the first half of 1902. Bulletin Number 37, new series, Division of Entomology, U. S. Department of Agriculture. November.
Osborn, Herbert. 1902c. A statistical study of variations in the periodical cicada. The Ohio Naturalist 3: 323-326.
Ostry, M.E. and N. A. Anderson. 1979. Infection of *Populus tremuloides* by *Hypoxylon mammatum* at oviposition sites of cicadas *Magicicada septendecim*. Phytopathology 69: 1041.
Packard, Alpheus S. 1873a. Third annual report of the injurious and beneficial effects of insects in Massachusetts. 20th Annual Report of the Secretary of the Massachusetts Board of Agriculture 3: 16-20.
Packard, Alpheus S. 1873b. American Naturalist 7: 536.
Packard, Alpheus S. 1890. Insects injurious to forest and shade trees. Fifth Report of the U. S. Entomological Commission, Washington 5: 95-97.
Pechuman, L. L. 1985. Periodical cicada—Brood VII revisited. Entomological News 96(2): 59-60.
Peck, C. H. 1879. Thirty-first Report New York State Museum of Natural History 31: 19, 20, 44.
Pettit, R. H. 1903. Mosquitoes and other insects of the year 1902. Special Bulletin Number 17, Michigan Agricultural Experiment Station, January.
Phares, Dr. D. L. 1845. Woodville, Mississippi Republican May 17.
Phares, Dr. D. L. 1858. Woodville, Mississippi Republican May 5.
Phares, Dr. D. L. 1873. Southern Field and Factory. Jackson, Mississippi April.
Phares, Dr. D. L. 1873. Southern Field and Factory. Jackson, Mississippi August.
Potter, Nathaniel. 1839. Notes on the *Locusta septentrionalis americanae decim septima*. J. Robinson. Baltimore, Maryland. 29 p.
Quaintance, Altus L. 1902a. The seventeen-year locusts; how the adults feed. Rural New Yorker July : 511.
Quaintance, Altus L. 1902b. On the feeding habits of the periodical cicada. Bulletin Number 37, new series. Division of Entomology, U. S. Department Agriculture, November: 90-94.
Quaintance, Altus L. 1902c. The periodical cicada or seventeen-year locust. Bulletin Number 87, Maryland Agricultural Experiment Station, November: 65-116.
Quaintance, Altus L. 1903. Entomological notes from Maryland. Bulletin Number 40, Division of Entomology, U. S. Department of Agriculture, August: 47.
Rathvon, S. S. 1869a. Hatching of the 17-year cicada. American Naturalist 3: 106.
Rathvon, S. S. 1869b. Cicada notes. American Entomologist 2: 51.
Rathvon, S. S. 1870. Periodical cicada not in Kruetz Creek Valley. American Entomologist and Botanist 2: 372.
Reichel, Reverend Charles. 1804. Some particulars concerning the locust of North America. Barton's Medical and Physical Journal 1: 52.
Reid, K. H. 1971. Periodical cicada: mechanism of sound production. Science 172: 949-951.
Riley, Charles V. 1865. Seventeen-year locust. Prairie Farmer 32: 127.
Riley, Charles V. 1866. Prairie Farmer 34: 136.
Riley, Charles V. 1868a. The 17-year cicada. Prairie Farmer 38: 2.
Riley, Charles V. 1868b. Prairie Farmer 38: 10.

Riley, Charles V. 1868c. Entomology. Prairie Farmer Annual (Number 2 for 1869): 30-41.
Riley, Charles V. 1869. The periodical cicada. First Annual Report on the Noxious, Beneficial and other Insects of the State of Missouri: 18-42.
Riley, Charles V. 1870. The periodical cicada alias the 17-year and 13-year locust. American Entomologist and Botanist: 211.
Riley, Charles V. 1872. Fourth Annual Report of the Noxious, Beneficial and otherInsects of the State of Missouri 4: 30-34.
Riley, Charles V. 1876. Periodical cicada, "17-year locust." New York Semi-Weekly Tribune. June 23.
Riley, Charles V. 1877a. Entomological notes. Transactions of the Academy of Science of Saint Louis 3: 217-218.
Riley, Charles V. 1877b. The periodical cicada. Western Farmer's Almanac for 1878: 48.
Riley, Charles V. 1880a. The 17-year cicada in Iowa. American Entomologist 3: 25-26.
Riley, Charles V. 1880b. Fungus in cicada. American Entomologist 3: 14.
Riley, Charles V. 1880c. The periodical cicada. American Entomologist 3: 172-173.
Riley, Charles V. 1881a. The periodical cicada alias "17-year locust." American Entomologist 25: 479-482.
Riley, Charles V. 1881b. The periodical cicada alias "17-year locust." Farmer's Review 6: 370.
Riley, Charles V. 1881c. *Cicada tredecim*, abundant in Alabama, as predicted. Selma, Alabama Times. July 19.
Riley, Charles V. 1881d. The periodical cicada. American Agriculture 40: 132.
Riley, Charles V. 1883. American Naturalist 17: 322.
Riley, Charles V. 1885a. Destroying cicadas. Rural New Yorker 44: 353.
Riley, Charles V. 1885b. Expected advent of the locust. Scientific American 52: 320.
Riley, Charles V. 1885c. The periodical or 17-year cicada. Harper's Weekly 29: 363.
Riley, Charles V. 1885d. The periodical cicada. An account of *Cicada septendecim* and its *tredecim* race, with a chronology of all broods known. Bulletin Number 8 [old series], Division of Entomology, U.S. Department of Agriculture (June 17). 46 p.
Riley, Charles V. 1885e. Notes on the periodical cicada. Science 5: 518-521.
Riley, Charles V. 1885f. Periodical cicada in Massachusetts. Science 6: 4.
Riley, Charles V. 1885g. The influence of climate on *Cicada septendecim*. American Entomologist 1: 91.
Riley, Charles V. 1885h. The song-notes of the periodical cicada. Science 6: 264-265.
Riley, Charles V. 1885i. Notes on the periodical cicada. Scientific American Supplement (495): 7905-7907.
Riley, Charles V. 1886a. The periodical cicada, etc. Report of the Entomologist, Annual Report of the U. S. Commissioner of Agriculture for 1885. June 8, 1886: 233-258.
Riley, Charles V. 1886b. Some popular fallacies and some new facts regarding *Cicada septendecim* L. Proceedings of the American Association for the Advancement of Science for 1885 (August) 34: 334.
Riley, Charles V. 1891. Periodical locusts. Scientific American May 16: 313.
Riley, Charles V. 1893. Periodical cicada. Science 16: 86.

Riley, Charles V. 1894a. Circular of the Division of Entomology, U. S. Department of Agriculture. 4 p.
Riley, Charles V. 1894b. The periodical cicada. Report of the Entomologist in the Annual Report, U. S. Department of Agriculture 1893: 204-205.
Riley, Charles V. and Leland O. Howard. 1888. The periodical cicada in 1888. Insect Life1: 31.
Riley, Charles V. and Leland O. Howard. 1889. The periodical cicada in 1889. Insect Life1: 298.
Riley, Charles V. and Leland O. Howard. 1893. The present year's appearance of the periodical cicada. Insect Life 5: 298-300.
Robinson, F. C. 1880. Seventeen-year cicada in Pennsylvania. American Entomologist 3: 178.
Rockwood, C. G., Jr. 1887. An insect fight. Science 10 (237): 94.
Russell, L. M. and M. B. Stoetzel. 1991. Inquilines in eggnests of periodical cicadas (Homoptera: Cicadidae). Proceedings of the Entomological Society of Washington 93: 480-488.
Rutherford, H. 1868. New York Semi-Weekly Tribune. June 27.
Sajo, Karl. 1899. Die siebzehnjahrige Cikade (*Cicada septendecim*). Prometheus 10: 388-393, 401-406.
Sandel, Andreas. 1906. "Extracts from the Journal of." Pennsylvania Magazine of History & Biography 30: 448-449.
Sanderson, E. D. 1898. Entomology. Country Gentleman July: 573-574.
Sanderson, E. D. 1901. Three orchard pests. Bulletin of the Delaware Agricultural Experiment Station, January, 1902 (December 1901) 53: 13-19.
Sanderson, E. D. 1902a. Report of the entomologist: Delaware Agricultural Experiment Station 1902: 137-139.
Sanderson, E. D. 1902b. Notes from Delaware. Bulletin Number 37, new series, Division of Entomology, U. S. Department of Agriculture, November.
Say, Thomas. The Complete Writings of Thomas Say on the Entomology of North America. Bailliere Brothers. New York, New York. 2 volumes.
Schmitt, J. B. 1974. The distribution of brood ten of the periodical cicadas in New Jersey in 1970. Journal of the New York Entomological Society 82: 189-201.
Schott, Fred M. 1946. 17-year cicada notes for 1945. Journal of the New York Entomological Society 54 (2): 167-169.
Schwarz, Eugene A. 1888. Cicadas at Fortress Monroe in June, 1886. Proceedings of the Entomological Society of Washington 1: 52.
Schwarz, Eugene A. 1890. Notes on *Cicada septendecim* in 1889. Proceedings of the Entomological Society of Washington 1: 230-231.
Schwarz, Eugene A. 1897. The periodical cicada in 1897. Division of Entomology, U. S. Department of Agriculture Circular, s. s. 22: 4.
Schwarz, Eugene A. 1898. The periodical cicada in 1898. Division of Entomology, U. S. Department of Agriculture Circular , s. s. 30: 4.
Shufeldt, R. W. 1894. The 17-year cicada and some of its allies. Popular Science News 28: 154-155.
Simmons, J. A., E. G. Wever, and J. M. Pylka. 1971. Periodical cicadas: sound production and hearing. Science 171: 212-213.

Simmons, J. A., E. G. Wever, W. F. Strother, J. M. Pylka, and G. R. Long. 1971. Acoustic behavior of three sympatric species of 17-yr cicadas. Journal of the Acoustical Society of America 49: 93.

Simon, Chris. 1979a. Brood II of the 17-year cicada on Staten Island: timing and distribution. Proceedings of the Staten Island Institute of Arts and Science 30 (2): 35-46.

Simon, Chris. 1979b. Evolution of periodical cicadas: phylogenetic inferences based on allozyme data. Systematic Zoology 28: 22-39.

Simon, Chris. 1983. Morphological differentiation in wing venation among broods of 13-year and 17-year periodical cicadas. Evolution 37: 104-115.

Simon, Chris. 1988. Evolution of 13- and 17-year periodical cicadas (Homoptera: Cicadidae: *Magicicada*). Bulletin of the Entomological Society of America 34: 163-176.

Simon, Chris and Monte Lloyd. 1982. Disjunct synchronic populations of 17-year periodical cicadas; relicts or evidence of polyphyly? Journal of the New York Entomological Society 90(4): 275-301.

Simon, Chris, Carl McIntosh, and Jennifer Deniega. 1993. Standard restriction fragment length analysis of the mitochondrial genome is not sensitive enough for phylogenetic analysis or identification of 17-year cicada broods: the potential for a new technique. Annals of the Entomological Society of America 86: 142-152.

Simon, Chris, Richard Karban, and Monte Lloyd. 1981. Patchiness, density, and aggregative behavior in sympatric allochronic populations of 17-year cicadas. Ecology 62 (6): 1525-1535.

Slingerland, M. V. 1893. The "17-year locust" in its hole. Rural New Yorker July 29: 509.

Slingerland, M. V. 1894a. The periodical cicada or locust. The Farmer's Advocate June 1: 225.

Slingerland, M. V. 1894b. The periodical cicada, or 17-year locust. Rural New Yorker July 28: 470 and August 4: 488.

Slingerland, M. V. 1896a. The apple crop and 17-year locusts. Rural New Yorker January 25: 53.

Slingerland, M. V. 1896b. On what do 17-year cicadas live? Rural New Yorker May 23: 351.

Slingerland, M. V. 1896c. Seventeen-year locusts not poisonous. Rural New Yorker July 11: 464.

Slingerland, M. V. 1897. Do 17-year locusts damage fruit trees? Rural New Yorker July 3: 437.

Slingerland, M. V. 1901a. Work of the 17-year locust. Rural New Yorker October 12: 690.

Slingerland, M. V. 1901b. Seventeen-year locust. Watermelon bug. Rural New Yorker July 13: 484.

Slingerland, M. V. 1902. Whitewash for 17-year locusts. Rural New Yorker March 15: 189.

Smith, Dr. Gideon B. 1851. The American locust "*Cicada septendecim*." Scientific American 1851: 212.

Smith, Dr. Gideon B. 1859. Country Gentleman 13: 308.

Smith, F. F. and R. G. Linderman. 1974. Damage to ornamental trees and shrubs resulting from oviposition by periodical cicada. Environmental Entomology 3: 725-732.

Smith, John B. 1889. The periodical cicada. Garden and Forest: 436.
Smith, John B. 1893. The periodical cicada. Bulletin Number 95, New Jersey Agriculture Experiment Station: 6.
Smith, John B. 1894a. The periodical cicada. Entomological News 5: 145.
Smith, John B. 1894c. The periodical cicada. Report of the Entomologist, Annual Report of the New Jersey Agricultural Experiment Station: 582-591.
Smith, John B. 1899. The periodical cicada. Report of the Entomologist of the New Jersey Agriculture College Experiment Station for 1898: 447-450.
Smith, John B. 1900. Notes on the occurrence of brood XX of the periodical cicada, *Cicada septendecim*, in Ohio in 1900. Entomological News: 638-640.
Smith, John B. 1889. The periodical cicada. Annual Report of the Entomologist [New Jersey] for 1889: 270.
Smith, K. G., N. M. Wilkinson, Kathy S. Williams, and V. B. Steward. 1987. Predation by spiders on periodical cicadas. (Homoptera: *Magicicada*). Journal of Arachnology 15: 277-279.
Snodgrass, Robert E. 1921. The seventeen-year locusts. Annual Report of the Smithsonian Institution 191: 381-409.
Snodgrass, Robert E. 1927. The head and mouth parts of the cicada. Proceedings of the Entomological Society of Washington 29 (1): 1-16.
Soper, R. S., A. J. Delyzer, and L. F. Smith. 1976. The genus *Massospora*, entomopathogenic for cicadas. Part II. Biology of *Massospora levispora* and its host, *Okanagana rimosa*, with notes on *Massospora cicadina*, of the periodical cicadas. Annals of the Entomological Society of America 69: 89-95.
Spence, R. H. 1851. Proceedings of the Entomological Society of London 1: 103-104.
Stannard, Lewis J., Jr. 1975. The distribution of periodical cicadas in Illinois. Illinois Natural History Survey Biological Notes (91): 3-12.
Stephen, F. M., G. W. Wallis, and K. G. Smith. 1990. Bird predation on periodical cicadas in Ozark forests: ecological release for other canopy arthropods? Studies in Avian Biology 13: 369-374.
Steward, V. B. 1986. Bird predation on the 13-year periodical cicada (Homoptera: Cicadidae: *Magicicada* spp.) in an Ozark forest community. M. S. Thesis. University of Arkansas at Fayetteville. 62 p.
Steward, V. B., K. G. Smith, and F. M. Stephen. 1988. Red-winged blackbird predation on periodical cicadas (Cicadidae: *Magicicada* spp.): bird behavior and cicada responses. Oecologia 76: 348-352.
Strandine, Eldon J. 1940. A quantitative study of the periodical cicada with respect to soil of three forests. The American Midland Naturalist 24 (1): 177-183.
Strecker, Hermann. 1879. The cicada in Texas. Science News 1 (16): 256.
Surface, H. A. 1905. The seventeen-year locust in Pennsylvania. Pennsylvania Department of Agriculture Monthly Bulletin of the Division of Zoology 3 (6): 174-175.
Surface, H. A. 1906. The cicada or seventeen-year locust in Pennsylvania. Pennsylvania Department of Agriculture Monthly. Bulletin of the Division of Zoology 3 (12): 369-377.
Thoennes, Gregory. 1941. Effects of injuries caused by the cicada, *Magicicada septendecim*, on the later growth of trees. Plant Physiology 16 (4): 827-830.
Toolson, E.C. and K. K. Toolson. 1991. Evaporative cooling and endothermy in the 13-year periodical cicada, *Magicicada tredecim* (Homoptera: Cicadidae).

Journal of Comparative Physiology B, Biochemical, Systemic, and Environmental Physiology 161: 109-115.
Turnipseed, George F. 1964. Chemical control of periodical cicada, *Magicicada septendecim*, on apples in North Carolina. Journal of Economic Entomology 57 (2): 295.
Walsh, Benjamin D. 1865. Practical Entomology 1: 18-19.
Walsh, Benjamin D. 1866. Practical Entomology 2: 33.
Walsh, Benjamin D. 1867. Practical Entomology 2: 56.
Walsh, Benjamin D. 1868a. The 17-year locust. Dixie Farmer June 11.
Walsh, Benjamin D. 1868b. Study of the periodical cicada. American Entomologist 1: 7-8.
Walsh, Benjamin D. 1870. American Entomologist October: 335.
Walsh, Benjamin D. and Charles V. Riley. 1868a. The sting of the 17-year cicada. American Entomologist 1: 36-37.
Walsh, Benjamin D. and Charles V. Riley. 1868b. The periodical cicada. American Entomologist 1: 63-72.
Walsh, Benjamin D. and Charles V. Riley. 1869a. Out of evil there cometh good. American Entomologist 1: 202.
Walsh, Benjamin D. and Charles V. Riley. 1869b. Belated individuals of the periodical cicada. American Entomologist 1: 217.
Walsh, Benjamin D. and Charles V. Riley. 1869c. Eggs of periodical cicada in savin-twig. American Entomologist 1: 228.
Walsh, Benjamin D. and Charles V. Riley. 1869d. The periodical cicada; our first brood established. American Entomologist 1: 244.
Walsh, Benjamin D. and Charles V. Riley. 1869e. Insects named. American Entomologist 1: 251.
Walsh, Benjamin D. and Charles V. Riley. 1869f. American Entomologist 1: 117.
Ward, Lester F. 1885. Premature appearance of the periodical cicada. Science 5 (123): 476.
Warder, R. H. 1869. Notes on the periodical cicada. It does oviposit in evergreens. American Entomologist 1: 117.
Webster, Francis M. 1889. An early occurrence of the periodical cicada. Insect Life II: 161-162.
Webster, Francis M. 1892. *Cicada septendecim*. Notes. Ohio Bulletin No. 45: 210.
Webster, Francis M. 1892. Modern geographical distribution of insects in Indiana. Proceedings of the Indiana Academy of Science: 81-89.
Webster, Francis M. 1897a. The 17-year locust in Ohio. Ohio Farmer May 20: 40.
Webster, Francis M. 1897b. The San Jose scale and the periodical cicada. Newspaper bulletin, Ohio Agricultural Experiment Station, November. 2 p.
Webster, Francis M. 1897c. Brood XV [V] of *Cicada septendecim* in Ohio. Canadian Entomologist October: 225-229.
Webster, Francis M. 1897d. The periodical cicada, or so-called 17-year locust in Ohio. Bulletin of the Ohio Agricultural Experiment. Station (87): 37-68.
Webster, Francis M. 1899a. Distribution of broods XXII [X], V [XIII], and VIII [XIV] of *Cicada septendecim* in Indiana. Proceedings of the Indiana Academy of Science 1899: 225-227.
Webster, Francis M. 1899b. Entomology. Ohio Farmer August 31: 152.
Webster, Francis M. 1900a. The 17-year locust in Ohio. Ohio Farmer July 5.

Webster, Francis M. 1900b. Notes on the occurrence of brood XX [VIII] of the periodical cicada, *Cicada septendecim*, in Ohio in 1900. Entomological News 11: 638-640.

Webster, Francis M. 1901. Report of the committee on entomology. Annual Report of the Ohio State Horticultural Society for 1900: 1-2.

Weed, C. M. 1885a. The coming locust plague. Prairie Farmer 57: 329.

Weed, C. M. 1885b. The 17-year locust. Prairie Farmer 57: 393.

Weed, C. M. 1888. Cicadas or harvest flies and beetles. Popular Gardener: 45.

Weiss, Harry B. 1953. The periodical cicada in New Jersey in 1953. Journal of the New York Entomological Society 61 (3): 159-161.

Westwood, J. O. 1839-40. Classification of insects 2: 4.

Wheeler, G. L., K. S. Williams, and K. G. Smith. 1992. Role of periodical cicadas (Homoptera: Cicadidae: *Magicicada*) in forest nutrient cycles. Forest Ecology and Management 51: 339-346.

White, JoAnn. 1981. Flagging: host defenses versus oviposition strategies in periodical cicadas (*Magicicada* spp., Cicadidae, Homoptera). Canadian Entomologist 113: 727-738.

White, JoAnn. 1973. Viable hybrid young from crossmated periodical Cicadas. Ecology 54 (3): 573-580.

White, JoAnn. 1980. Resource partitioning by ovipositing cicadas. American Naturalist 115 (1): 1-28.

White, JoAnn, P. Ganter, R. McFarland, N. Stanton, and Monte Lloyd. 1983. Spontaneous, field tested and tethered flight in healthy and infected *Madicicada septendecim* L. Oecologia 57: 281-286.

White, JoAnn and Monte Lloyd. 1975. Growth rates of 17- and 13-year periodical cicadas. The American Midland Naturalist 94 (1): 127-143.

White, JoAnn and Monte Lloyd. 1979. 17-year cicadas emerging after 18 years: a new brood? Evolution 33 (4): 1193-1199.

White, JoAnn and Monte Lloyd. 1981. On the stainability and mortality of periodical cicada eggs. The American Midland Naturalist 106 (2): 219-228.

White, JoAnn and Monte Lloyd. 1983. A pathogenic fungus, *Massospora cicadina* Peck (Entomophthorales), in emerging nymphs of periodical cicadas (Homoptera: Cicadidae). Environmental Entomology 12 (4): 1245-1252.

White, JoAnn and Monte Lloyd. 1985. Effect of habitat size on nymphs in periodical cicadas (Homoptera: Cicadidae: *Magicicada* spp.). Journal of the Kansas Entomological Society 58: 605-610.

White, JoAnn, Monte Lloyd, and Richard Karban. 1982. Why don't periodical cicadas normally live in coniferous forests? Environmental Entomology 11 (2): 475-482.

White, JoAnn, Monte Lloyd, and J. H. Zar. 1979. Faulty eclosion in crowded suburban periodical cicadas: populations out of control. Ecology 60 (2): 305-315.

White, JoAnn and C. E. Strehl. 1978. Xylem feeding by periodical cicada nymphs on tree roots. Ecological Entomology 3: 323-327.

Williams, Kathy. S. 1987. Responses of persimmon trees to periodical cicada oviposition damage. Pages 424-425 *in* Insects-Plants. V. Labeyrie, G. Fabres, and D. Lachaise, editors. Dordrecht: Junk. 459 p.

Williams, Kathy S. and Chris Simon. 1995. The ecology, behavior, and evolution of periodical cicadas. Annual Review of Entomology 40: 269-295.

Williams, Kathy S. and Kimberly G. Smith. 1991. Dynamics of periodical cicada chorus centers (Homoptera: Cicadidae: *Magicicada*). Journal of Insect Behavior 4: 275-291.

Williams, Kathy S. and Kimberly G. Smith. 1993. Host plant choices of periodical cicadas, *Magicicada tredecassini*. Bulletin of the Ecological Society of America 74 (2): 489 (Supplement).

Williams, Kathy S., Kimberly G. Smith, and F. M. Stephen. 1993. Emergence of 13-yr periodical cicadas (Cicadidae: *Magicicada*): phenology, mortality, and predator satiation. Ecology 74: 1143-1152.

Woodworth, C. W. 1888. Synopsis of North American Cicadidae. Psyche 5: 67-68.

Young, D. and R. K. Josephson. 1983. Pure tone songs in cicadas with special reference to the genus *Magicicada*. Journal of Comparative Physiology A, Sensory, Neural, and Behavioral Physiology 152: 197-207.

Young, Frank N. 1958. Some facts and theories about the broods and periodicity of the periodical cicadas. Proceedings of the Indiana Academy of Science 68: 164-170.

Young, Frank N. 1971. Observations on periodical cicadas (brood X) in Indiana in 1970 (Homoptera-Cicadidae). Proceedings of the Indiana Academy of Science for 1970 (80): 247-252.

Young, Frank N. 1975. Observations on periodical cicadas (brood XIV) in Indiana in 1974 (Homptera-Cicadidae). Proceedings of the Indiana Academy of Science 84: 289-293.

Young, Frank N. 1977. Observations on periodical cicadas (brood XXIII) in Indiana in 1976. Proceedings of the Indiana Academy of Science for 1976 86: 244-245.

Young, Frank N. and Gene Kritsky. 1987. Observations on periodical cicadas (brood X) in Indiana in 1987 (Homoptera: Cicadidae). Proceedings of the Indiana Academy of Science 97: 323-329.

# Notes